Eating the Ocean

Eating the Ocean

ELSPETH PROBYN

Duke University Press   Durham and London   2016

© 2016 Duke University Press
All rights reserved
Printed and bound by CPI Group (UK) Ltd, Croydon, CR0 4YY
Cover design by Heather Hensley; interior design by Courtney Leigh Baker
Typeset in Whitman by Graphic Composition, Inc., Bogart, Georgia

Library of Congress Cataloging-in-Publication Data
Names: Probyn, Elspeth, [date] author.
Title: Eating the ocean / Elspeth Probyn.
Description: Durham : Duke University Press, 2016. | Includes bibliographical references and index.
Identifiers:
LCCN 2016018768 (print)
LCCN 2016019478 (ebook)
ISBN 9780822362135 (hardcover : alk. paper)
ISBN 9780822362357 (pbk. : alk. paper)
ISBN 9780822373797 (e-book)
Subjects: LCSH: Food habits—Environmental aspects. | Sustainable fisheries. | Seafood—Environmental aspects. | Seafood industry—Environmental aspects. | Feminist theory.
Classification: LCC GT2850.P76 2016 (print) | LCC GT2850 (ebook) | DDC 333.95/616—dc23
LC record available at https://lccn.loc.gov/2016018768

Cover art: Giant bluefin tuna. Photo by Jeff Rotman / Alamy.

CONTENTS

Acknowledgments vii

Introduction RELATING FISH AND HUMANS 1

1 An Oceanic Habitus 23

2 Following Oysters, Relating Taste 49

3 Swimming with Tuna 77

4 Mermaids, Fishwives, and Herring Quines
  GENDERING THE MORE-THAN-HUMAN 101

5 Little Fish EATING WITH THE OCEAN 129

Conclusion REELING IT IN 159

Notes 165
References 169
Index 183

ACKNOWLEDGMENTS

The genesis and the development of this book came from many travels and conversations, and I thank all those who shared their knowledge, stories, and love of fish and oceans with me. This project allowed me to discover an immensely complex and generous world. I discovered bluefin tuna and the idea of researching human-fish communities while I was at the University of South Australia, and I thank the former vice chancellor, Peter Høj, for generously funding me, and the great gang at the Hawke Research Institute (Lyn Browning, Gilbert Caluya, Maureen Cotton, Sonia Saitov, Lisa Slater, and Shvetal Yvas). Thanks to Mandy Thomas for her support over glasses of wine in Adelaide. Funding from the Australian Research Council is gratefully acknowledged. My wonderful colleagues and our students at the Department of Gender and Cultural Studies drew me back to the University of Sydney, and I thank them for their continuing generosity and intellectual support. It is my intellectual home, and I feel very lucky to have such wonderful people in my life. Kate Johnston and Nancy Lee helped me with research and, along with Morgan Richards's wonderful graphic skills, were central to the development of the Sustainable Fish Lab. Sean Fuller is a wizard editor. Jodi Frawley shared her love and knowledge of fish. Katrina Schlunke prodded me along when I was drowning in fish facts. Clif Evers, man of the sea, reminded me of the importance of tacit knowledge. Jennifer Biddle (cherished book-writing buddy) and I swapped chapters to read, and we cheered each other along at the Angry Pirate. I have gained much from the many engaged audiences who have responded to papers I have given along the way, and my thanks to them and to the organizers who invited me. Thanks to Ken Wissoker, who thought this book was a good idea, and to Elizabeth Ault for her careful shepherding.

My deepest thanks and love go to my partner, Sarah Donald, who was there at every turn of our fish travels, endlessly driving us around Australia, Scotland, and the Pacific Northwest chasing fish, taking photographs, formatting the images in the book, finding out interesting fish details, and eating fish, seafood, and oysters with me.

I dedicate this book in loving memory to my father, Alfred John Probyn, fisherman and raconteur extraordinaire. My late brother, Stephen Probyn, loved tuna before I encountered them. Both are greatly missed by my sister, Jane Probyn, and me.

Introduction RELATING FISH AND HUMANS

Imagine a sparkling day on Sydney Harbour . . . eating creamy Sydney rock oysters and drinking a glass of fine Australian white wine with a light warm salty breeze on your face. Look—there's the Opera House with her shells; to your left is the Sydney Harbour Bridge (the Coathanger) that each New Year's Eve features actual over-the-top fireworks touted as the first to be seen globally on the first day of a new year. On the water there's the crazy skittering of the ferries, sailboats, and kayaks crisscrossing east and west, north and south. Around Circular Quay the usual huddles of tourists handle the faux UGG boots and the didgeridoos made in China. Next to the berth for the ferry to Manly, Aboriginal musicians play proper didge. Over by the old shipping wharves—now eye-wateringly expensive real estate—some young boys but mainly old women and men fish. Many came from Vietnam on boats years ago, and fresh fish still feeds the family. Day in, day out, they sit on milk crates fishing under the bridge. Holding all these stories together is the water of Sydney Harbour—it is normally a color called harbor green, but sometimes it burnishes to a shimmering near-turquoise. There's something like five hundred gigaliters of water in the harbor, an amount that is called one Sydharb. Below the surface swim some 586 different species of fish. In among the local fish there are now tropical fish who, like the clownfish in *Finding Nemo*, ride the East Australian Current over a thousand miles down from the Great Barrier Reef in the north.

It sounds rather magical, and it often is, at least on a surface level. The reason why you can now go snorkeling in the harbor and encounter tropical fish is that this part of the Pacific Ocean is warming faster than anywhere else in the world. And unbeknownst to many, the water in the harbor leads the world in the amount of heavy metals it contains. The sludge at the bottom is anoxic slurry with no oxygen for life. Much of it settled in the seabed when it was acceptable to throw industrial waste into the harbor thirty years ago. But dioxins are continually flowing in the storm waters that usher the poisons of the city into the sea. Every year, five hundred gigaliters—one Sydharb—of storm water gushes untreated into the harbor. Commercial fishing was banned in 2006 because of the toxicity, and recreational fishers are warned not to eat more than minuscule amounts of fish caught to the east of the bridge, and to never eat fish caught to the west. This divide strangely maps onto another: The wealthy tend to live in the east of Sydney, with the poorer to the west. The official site of the NSW Department of Primary Industries gives advice to recreational fishers in seven languages. It says fishing licenses are mandatory, but it doesn't tell people what may happen to their bodies if they eat the fish they catch.

Some might call the harbor a man-made cesspit. But it's got a lot of life in it—for now. Emma Johnston and her team at the Sydney Harbour Research Program are finding that fish are flourishing. After two years of research, this large interdisciplinary team of scientists reports that many species are coping rather well with the anthropogenic modification of their ecosystem. It is a paradox: "Increased nutrient levels may be enhancing the productivity levels of the system and hence the abundance of fish" (McKinley et al. 2011, 643). The nutrients are mainly due to the nitrates and pesticides that are carried in the storm water directly into the harbor. Fish get used to the man-made modification of their world and take advantage of the increased nutrients. They seemingly thrive, and yet they are poisonous for their human and nonhuman predators.

Eating the ocean: We do it every day, often without knowing it. Humans have eaten the ocean for as long as we've been around. Until relatively recently we thought that we could eat it with impunity. Now we are at risk of eating it up, devouring it until there's nothing left except the not-so-apocryphal jellyfish-and-chips. "We" are, of course, differently positioned in this scenario.

This book is an intervention into the current politics of food, although they are still by and large concerned with terrestrial production of food. Along with others, I'd say that some of the academic debates and media discussions about alternative food practices have tended to become rather simplified, especially when they are fixated on urban localism. The mantra "local is best" barely hides its white middle-class complexion. And a certain moralism pervades the discourse about what is good to eat, and why some people eat badly, that is often paired with the desire to enlighten the unenlightened about their bad food habits. In Australia, as elsewhere, this is often racialized. "Why are Aboriginal and Torres Strait populations three times more at risk of developing type 2 diabetes than non-Indigenous Australians?" "Why don't Aboriginal people eat better?" The answers to such questions are economic, cultural, and historical. Fresh food in remote Australia is many times the price that it is in the urban south, a situation that some non-Indigenous Australians cannot countenance in their depiction of Aboriginals refusing to eat better. Underlying this is a deep and tragic history whereby white colonists bribed Aboriginal people with white sugar and white flour—what Tim Rowse (2002) terms "white death."

Food, as many have pointed out, is far from simple. The rhetoric of helping people to eat better is drenched in condescension. Julie Guthman's (2007) trenchant critiques highlight the ways in which the mantra of "bringing good food to others" and the appeal to the inherent good of organic food forgets how historically organic food production was wedded to eugenic desires for racial and nutritional purity. In the context of the United States, Guthman also notes that localism, the bedrock of alternative food rhetoric, can be xenophobic and historically blind. For many, "the local" was not a romantic ideal: It was where people of color or people marked by class were scrutinized and shamed. This scrutiny often continues in the snide glances directed at working-class women with shopping carts loaded with cheap carbohydrates to fill up families on a meager wage or pension.

Guthman notes "the extent to which food politics have been at the cutting edge of neoliberal regulatory transformations" (2007, 437). While critiques of neoliberalism often leave me unimpressed, there are certainly places where neoliberal food regimes are in full force—such as food stamp programs or the Basics card forced upon Aboriginal populations in the Northern Territory. Then there is the multivalent and contradictory way in which choice operates in some forms of food politics. As with the condescending attitude toward those who don't choose to eat better, increasingly the choice to proclaim oneself vegan often seems to act as an opting out of the structural complexities of food provisioning, production, and consumption.

Considering food through the optic of fish considerably complicates food debates. Against the mantra of the local, it is nigh impossible, given the arrangements of fishing industries, to eat a local fish. Sure, you can sometimes catch your own (though not in Sydney Harbour if you want to stay healthy), or very occasionally you can find a fisher at a pier who can flip you a freshly caught fish, or you can poach—as in steal—a salmon caught in a Scottish river. The assumption that a shorter supply chain results in more ethical food is considerably problematized in fishing. Because of the restructuring of international fishing practices brought on by the necessity of regulating catch, in the last twenty years the size of fishing fleets in the Global North has shrunk to a fraction of what it was. The use of individual transferable quotas (ITQs), introduced in Iceland, Canada, and Australia in the late 1970s and early 1980s and later adopted elsewhere, stopped the practice of fishing in common. While I go into detail on the practice of ITQs later in the book, the salient point here is that the introduction of quotas quickly sliced into the number of boat owners, resulting in the current state where the ownership of fishing is in a few hands. For instance, in the South Australian bluefin tuna industry, the number of licensed fishing boat owners went from several hundred to under thirty as soon as the ITQ was introduced. Concomitantly, the downturn in inshore fishing because of overfishing, and the need to amortize the costs of increasingly sophisticated technology to track fish, has led to ever-larger boats that can go further out. Simply put, this means you cannot go down to the dock and "look the fisherman in the eye" (pace Michael Pollan's dictate to "look the farmer in the eye"). The long sea trips and the widespread use of freezing technology means that fish are caught and flash frozen out at sea. They are then landed and immediately transferred to large logistical operations that take the fish hundreds or several thousands of miles away to market. As I discuss later in the book, some operations do try to go against this structuring of the industry. The community-supported fisheries on the northwestern Pacific Coast in the United States operate like community-supported agriculture, and consumer-members receive boxes of fish, thus providing a regular income to fishers and encouraging people to be more adventuresome in the range of fish they consume. The ThisFish operation in Canada links fishers, suppliers, and consumers via social media—including a photograph of the fisher so you can nearly look him in the eye. My point here is that fish-as-food requires us to go beyond a simplistic food politics. It compels us to understand how entangled we are as consumers in the geopolitical, economic, cultural, and structural intricacies of the fishing industries.

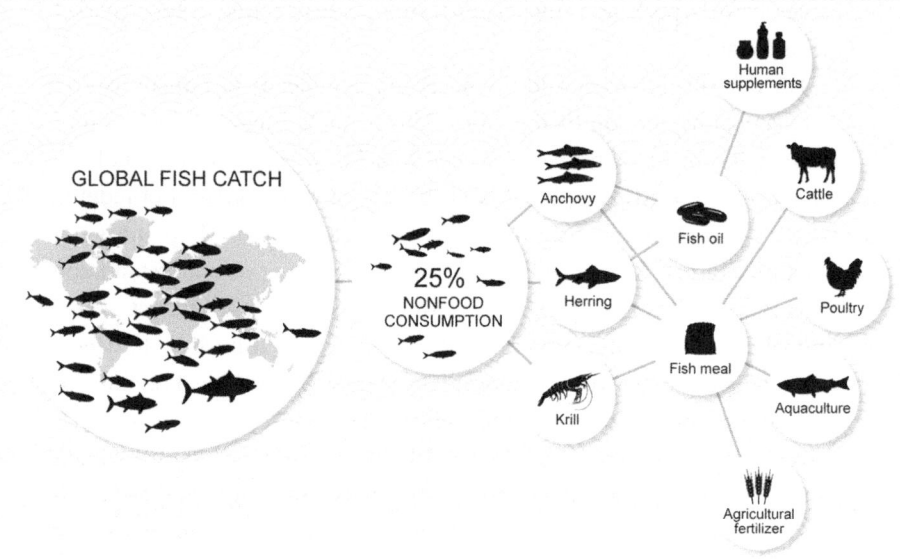

Fig. 1.1. Fish as food. Illustration by Morgan Richards.

Fish-as-food recalibrates the extent to which anyone can choose to opt out of the dominant food regime by saying, "I don't eat fish." As it stands, we all eat fish, albeit often in circuitous ways. It is staggering where and how fish enter the food chain without being fish for human consumption. As figure 1.1 shows, 25 percent of the global fish catch ends up in strange places: as fertilizer, as food for pigs and poultry, as fish oil supplements, and, perhaps most ghoulish of all, as processed food for fish.

My point is simple: There is no innocent place in which to escape the food politics of human-fish entanglement. Another point follows from the fact that localism cannot reassure us: Human-fish entanglements are not linear. They are simultaneously global and local, regional and hemispheric, Global North and South. This means that a food politics cannot start with the fisher and follow through to the consumer. We have to understand fish-as-food from the middle of the complex entanglement of industries, of historic and colonial trade routes that persist in the strip mining of African fisheries by the Global North, of the producers who may be iconic rugged individual fishermen or who may be indentured labor on shrimp farms in Southeast Asia. More often than not, your fish will be fed by transient flows of capital, labor, and intensive technology. Fish may also feed your vegetables, cereals, pork, and chicken. Fish are even to be found in supermarket white bread.

For me, this intricacy and the necessary complexity of human-fish relations ups the ante on food politics. This is not academic obfuscation or complexity for its own sake. We need to acknowledge that food politics have become overly simplified, and that the stoplight system of judging what's a good fish or a fish to avoid is seriously limited. My point is not pedantic; rather, I hope to demonstrate that coming to understand and explore fish-human complexity is exciting—it compels our interest in how and where we are integrated into this watery world. It leads us to fishers and producers, to networks of movement around the world that reignite for me the thrill of thinking global movement. The research for this book has moved me in many ways—the loss of fish, habitat, and the tacit knowledge embedded in many fishing communities makes me angry and sad. This research has also moved me away from much of my previous work that was focused on overtly cultural matters of eating—how, for instance, our identities are forged through eating. In publications such as *Carnal Appetites* (Probyn 2000), I was mainly concerned with a more abstract contemplation and depiction of the ethics of eating. For me, food really was good to think with, and my analysis tended to float away from the specificity of how, where, and by whom food was produced, and with what effects.

In 2009 I was hired by the University of South Australia to help energize their research. I'm not sure I did much good, but I certainly learned a lot. Beyond its capital, Adelaide, South Australia is a beautiful, sparsely populated, and isolated state. Economically it has been based in natural resources: wheat, sheep, wine, mining, and fish. Working with rural sociologists, I became interested in the communities involved in primary food production. We called this research program Producing Identities.[1] South Australia was then in deep drought—the driest state on the driest continent. The Murray River, the only river in the state, was being drained away through drought and irrigation. Times were truly tough. The range of issues was and is complex: the ethnic mix in farming communities, the social challenges of balancing the human need for scarce fresh water with the complex other-than-human ecologies that depend on it for their lives. Then there's the backdrop of depression in farming communities and the high rates of suicide, especially among male farmers.

A colleague suggested that I visit the small fishing communities along the coast of the Eyre Peninsula.[2] I fell in love with fish. I also became entranced with the intricate forms of biocultural sociality that weave through small communities that depend on fishing and oyster farming.

This book comes out of that initial experience, which has developed as a larger project.[3] But it also comes out of a lifelong concern about how we can ensure global and equitable access to protein. For many years, I was a fish-eating vegetarian. Given that the earth is 70 percent ocean, I thought eating fish and not eating terrestrial beings could help with the global distribution of protein. It is estimated that the food conversion rate for beef is between eight and twenty pounds for a pound of beef. Thirty-eight percent (about 730 million tons) of the world's grain harvest is used to produce animal protein (Brown 2006). Conversely, advances in technology mean farming salmon can require as little as two pounds of feed to produce a pound of protein. Of course some of that feed comes from fish. The case for aquaculture is not straightforward, and, as I discuss later, there are several practical and ethical problems with fish farming. But from the experience of seeing up close how dry a drought is—think kiln-baked soil—I know that we cannot continue to feed people only through terrestrial means. And by people, I mean the estimated future nine billion humans. To feed that population, we would need to increase cropping and grazing by 70 percent. That doesn't factor in the water, the paucity of arable land, the fossil fuels, peak phosphorus, and the damage done by pesticides to people, ecosystems, and the sea. Should we then be asking, "Will the oceans help feed humanity?" (Duarte et al. 2009).

Eating the Ocean responds to the challenge of how to produce and consume fish in a sustainable way. The term "sustainability" is, of course, problematic. It covers so much that it becomes nigh on meaningless. It may conjure feel-good affects, but materially it continues to be intractable. UNESCO's four-pillar model of sustainability adds culture to the economic, the social, and the environmental, which is better than ignoring culture altogether. But often culture is sequestered, confined to the folkloric.[4] In addition, the iteration of what is supposed to be sustainable fish evinces little concern or interest in what it would take to sustain the biocultural relatedness of fish and humans that is millennia old. Again and again through this book, I wonder how we can care a bit more, or a bit better, for the entire entangled marine elements that we devour when we eat the ocean. I also ask—as in the title of the last chapter—can we eat *with* the ocean?

This book is deeply invested in teasing out the very different sorts of knowledge that construct what eating the ocean means. There are disparate and competing models of what fish are and what they can and should be in the future. Scientific and ecological models insist that numbers, quantity, and eco-sustainability are what matter; local fishing communities value economic viability, traditional practice-based knowledge, and ways of life; consumers operate in budget-defined regimes of commodified taste and choice. Each of these spheres brings to the table radically different models of what fish represent and what they do in deeply implicated, mutually dependent human-fish networks. Groups of people and sets of meaning are all too often rendered separate, if not antagonistic: fishery managers versus fishers, conservationist groups versus fishers, and even fishers versus fishers.

What is clear is that there are less and less fish on the table. If we take the figures from the United Nations' Food and Agriculture Organization (FAO),

Of the 600 marine fish stocks monitored:
3% are underexploited
20% are moderately exploited
52% are fully exploited
17% are overexploited
7% are depleted
1% are recovering from depletion[5]

These figures raise more questions than they resolve. The definition of "exploitation" is simple yet baffling: The FAO states that "overexploited" means "the fishery is being exploited at above [sic] a level which is believed to be sustainable in the long term." Defining "exploitation"—hardly a reassuring term—rests on maximum sustainable yield, the keystone of a fisheries management formula developed in the 1930s. This is the mathematical formula that calculates the amount of fish harvested against the population of the species, with an estimate of the rate at which the species reproduces. The problem is that it is very hard to count fish. As one of my interviewees puts it, they have tails. They swim across artificial lines in the sea. Maribus, a German consortium of scientists, depicts how the FAO comes to formulate these figures: "The catch data from both the fishermen and the scientists is initially forwarded to higher scientific institutions which utilize it to estimate the current stocks of the various fish species and maritime regions. Around 1500 fish stocks around the world are commercially fished, with the various stocks being exploited to different extents. Comprehensive estimations of abundance currently exist for only around 500 of these stocks"

(World Ocean Review 2015). It's often said that we know more about the moon than we do about what is in the oceans. Fishery scientists just do not know how many species are in the ocean, nor indeed much about their behavior. The case of the orange roughy (renamed from its original appellation of slimehead for obvious reasons) is a stark example. Exploitation quotas for this deep-sea fish were based on the assumption that it had a short life span and reproduced quickly. By the time it was apparent that an orange roughy could live up to 150 years, and that it only reproduces when it is about thirty years old, the roughy population was devastated.

The FAO figures that are the cornerstone of fisheries management rely on numbers compiled by member nation-states. But there are few accurate historical records to tell us how the fisheries are faring. In the early 1990s, Daniel Pauly and his team pointed out that there was no adequate understanding of what constitutes a baseline for fish stocks (Pauly 1995). To make things worse, sometimes countries simply lie about their catch numbers. In 2001, Pauly and his colleague Reg Watson reported that China systematically inflated its catch numbers, leading to the impression that global fishing catches were fine. Now the FAO excludes the Chinese catch because it skews the picture so badly (Watson and Pauly 2001).

We know that numbers mean nothing without an explanation of what is counted, how, and by whom. The immense fragility of human knowledge that is the scaffold for gauging sustainability is well known among scientists and fisheries officials, yet seldom remarked upon in public. Who does the fishing is equally obscure to the public. Despite the advertising images of John West and jolly old sea captains, women are essential to fisheries—the FAO (2015a) estimates that women account for at least 15 percent of people directly engaged in the fisheries primary sector, and 90 percent in the secondary sectors such as processing. Those figures don't really tell us much. For instance, in the massive Thai shrimp industry that produces the bulk of the shrimp eaten in the United Kingdom and northern Europe, most of the workers are female—often illegal—migrants, who are not paid a living wage, and in some cases are considered indentured labor working off the costs of getting to Thailand for a poorly paid job (Fairfood 2015).

Who tells the stories, how, to whom, and why is a theme that reverberates through this book. I continually shadowbox with several dominant discourses that reduce the complex worlds of fish, oceans, and humans to simple black-and-white distinctions. There is ample detail in the following chapters, and here I again simply state that saying no to fish is not an option. We will see that there are times and places and certain fish that some humans in some places should not envision eating. But as I have flagged already, I argue strongly against the hubris that passes for a politics of fork waving. The idea that you can resolve such intricate and complicated human-fish relations by voting with your fork is deluded narcissism.

Three main currents organize this book. These currents draw together stories, fieldwork notes, arguments, and ideas into eddies that swirl in and across the book: (1) necessary complexity; (2) ethologies of the more-than-human; and (3) gender and queer fish relations.

Necessary Complexity

The ocean, fish, and humans are all incredibly complex entities. It's hard enough to detail each one separately. Of necessity, the three spheres have to be continually interrelated. The scale of their entanglement is mind-boggling. In marine science, "simplified sea" refers to what happens when we fish down the food web, resulting in an ocean stripped of biodiversity. On land, the ocean is simplified in many ways: Authorities divide people on the basis of their knowledge or putative lack thereof. I've attended numerous meetings in local communities about impending large changes to the fishing environment where people's experiences and knowledge are categorized and then excluded. This makes the drawing of lines easier: the lines that divide the oceans into manageable blocks that then "belong" to certain groups.

I mobilize the idea of the more-than-human across this book. While it has its problems, one of the reasons that I like the term is that it compels relational thinking. Logically, at least, you cannot engage in polemic if your objective is to find, forge, and relate connection and complexity between and among human communities and marine ecologies. There are of course many versions of the more-than-human that I recount in the book. The term itself isn't that important, nor is it innocent. Who has the capital to blithely step in to the more-than? What of the many who still aspire to the status of the human? As Karen Cardozo and Banu Subramaniam argue, "the turn to including nonhuman animals in intellectual inquiries does not necessarily deconstruct a hierarchical Great Chain of Being" (2013, 1). If it does direct us to the complex and deeply unequal distribution of matter in which we are always differentially related, then all to the good.

When it doesn't work, it will need to be prodded and maybe refigured in other ways.

Annemarie Mol and her Eating Body team at the University of Amsterdam argue that all matter is matter related, which then compels "other modes of *doing*, such as affording, responding, caring, tinkering, and eating" (Abrahamsson et al. 2015, 6).[6] For them, matter is always "enmeshed in a variety of relations" (10). Some of the entities to which or in which fish are related are the environment, climate change, things that eat fish and fish that eat things, laws and regulations, technologies, and markets. These relations make for complex interactions. "All of which suggests that, rather than getting enthusiastic about the liveliness of 'matter itself,' it might be more relevant to face the complexities, frictions, intractabilities, and conundrums of 'matter in relation'" (13).

Wet Ethnographies and Ethologies: Relating

Eating the Ocean is necessarily complex in its subject matter, but I hope it doesn't read as convoluted. Let me quickly sketch the import of each chapter and how each may engage the other. In chapter 1, my focus is to set the scene, to move us away from terra firma to the many affective realms of the sea. Acknowledging our human difference, and perhaps the ocean's indifference to human life, is a starting point in a journey to make strange the relations of human, oceans, and fish. I start by deepening my critique of predominantly land-based food politics, and the emerging nongovernmental organization (NGO) campaigns against fishing. I hope that by shifting our attention offshore, new angles of intervention become possible. I start to explore a thread that builds across the book, patching together an argument, a vision, a hope for an affective oceanic habitus. In chapter 2, I turn to smaller marine ecosystems and the different forms of attachments and metabolic intimacy that are forged by oysters—a marvelous sustainable and hardworking marine entity that is also delicious to eat. At one register I take from Annemarie Mol's argument about what happens when we consider seriously the interaction of eater and object, a focus that mires any straightforward distinction of agency and that otherly conjures relations of subjectivity. If Mol (2008) rehearses what happens when "I eat an apple," in my rendition, when "I eat an oyster," it also becomes clear that "an oyster eats me." Oyster eating is a rare instance when live flesh meets live flesh. On another register, the massive business of farming predominantly Pacific oysters in different parts of the world forms global networks of oyster-human communities. I follow oysters to Loch Fyne, a small village in the west of Scotland, and then to a small community in South Australia where remarkable entanglements of history, care, oysters, and humans have produced a flourishing more-than-human culture. In yet another take on oyster-human relatedness, I draw on marvelous raconteurs of oysters such as M. F. K. Fisher (1990) and her odes to the oyster, "Love and Death among the Molluscs" and "Consider the Oyster." Through these seemingly disparate elements, I relate how the mattering of oysters highlights taste as a simultaneously economic, cultural, and more-than-human affair. In chapter 3, I dive into the contested waters of bluefin tuna consumption. As I learn to swim with tuna, I also come to appreciate the relations of technology, ethnicity, markets, and geography that have rendered the magnificent bluefin tuna into a breathtakingly expensive commodity. In this chapter, I follow and relate the stories of how this came to be. Swimming alongside these stories, I come to realize a different ethical take on why we shouldn't eat bluefin tuna.

In chapter 4, I plunge into a lacuna that marks much of the research on the more-than-human. In the age of the human, gender seems to be passé. From a discussion of mermaids through to the lives of fisherwomen, I seek to elevate gendered and queer matters of human-fish entanglement. At a conceptual level, I grapple with how gender and sexuality, as well as ethnicity and class, have been squeezed out of current debates on the Anthropocene, climate change, and the more-than-human. There is a pervasive sense that the big issues of the Anthropocene override the concerns of feminists, queers, and postcolonial people, and the questions of race and class. The idea that we are all brought together, that our differences are elided by living under the shadow of the Anthropocene, is, of course, nonsense. We are not nor will we be all equally affected by the multiple disasters occurring within the rubric of the man-made Anthropocene. To be blunt, in some circles the Anthropocene licenses a focus on the human, who often turns out to be male. But rather than mounting a polemic, in this chapter I seek to recover the lost stories of the women who have followed fish, and see what happens to human-fish settlements when the fish disappear. Two instances of this ground the chapter in history—the rise and fall of the herring industry in Scotland that lasted from the nineteenth century until after World War II, and the collapse of the cod fisheries of the Great

Banks off Newfoundland and Labrador in Canada. In the case of the former, I gather the stories and depictions of the herring quines—the lassies who followed the herring from the very north of Scotland to the south of England, packing some thirty thousand herrings a day. These stories are gathered through historical documents and encounters with the women and men who across the generations remember them. In the latter, when it was announced in 1992 that the cod fisheries in Canada were to be closed (the moratorium), fifty thousand workers were rendered redundant. Attention turned to the plight of the male fishers. Feminist sociologists such as Barbara Neis, Donna Lee Davis, and Nicole Gerarda Power intensively researched the fishing communities before, during, and after the moratorium. They relate the insights of the women who worked as processors or who as fisher wives did the accounts of the family fishing business. These women attest to the dwindling of the stock long before the closing of the fishery. They were not listened to. Perhaps even more galling is that in the aftermath of the crisis, their insights about how to better manage the fishery went unheard. This chapter is, I hope, a salutary reminder of the varied types of gendered tacit and contextual knowledge that must be taken up if we are to respond to what Daniel Pauly (2009) calls the "Aquacalypse."

In the last chapter, it is little fish that come to the surface. In my research for this book, fisher-people whom I talked to often spontaneously remarked: "These are the fish I love." Anchovies, sardines, herrings, and menhaden—these are the fish I particularly love, along with even smaller marine organisms such as algae and various forms of phytoplankton. When people ask me what fish they should eat, I reply, "Little fish." They are the ones that reproduce quickly. But they are also often the fish that are reduced to mere fodder. In this chapter we follow little fish into other more positive entanglements, such as the emerging technologies of integrated marine trophic aquaculture (IMTA). These practices take from ancient practices of polyculture in Asia and elsewhere (e.g., a duck on a rice paddy whose feces feed the fish that swim underneath). In China they are now mammoth affairs where feed trickles down to fish and the nitrate-laden debris is eaten by bivalves and continues down to sea cucumbers and algae. Here nothing is wasted, as it all becomes forms of protein for humans and for marine organisms. This system may eventually feed humans with less impact on the ocean and her inhabitants. Smaller IMTA operations are under way across the world, helping to make aquaculture a more sustainable affair.

Eating the Ocean builds on my case studies of human-ocean-fish entanglement. They flow and move across the book and defy the somewhat linear chapter descriptions above. Oysters, bluefin tuna, fish-women and herring quines, sardines and anchovies—these sites, or entanglements, are both physical and at times troubling eddies where theories and empirical material come together in unexpected and clashing ways. The sites may seem arbitrary, but it is their capacity to express relatedness that draws me to them. The physical and geographical locations are storied places, which I then amplify with more stories. For instance, in chapter 2, Loch Fyne becomes a place where oysters bring together stories, history, and geographies. Histories of people are written into the water. As Raymond Williams (1989) would say, people live this relatedness, just as they live their "culture [as] ordinary." As a wet ethnographer—wet in the doubled sense of being a soft ethnographer who dredges oceanic tales—I tease out connections and relate them. Just as the practice of wet ethnography has a double valence, so too does relating. I relate the stories I'm told, and I relate their tales to theoretical and political concerns. As much as I can, I try to inhabit these relations, to make these acts of relating fleshy and fishy.

This is a dialogic and embodied practice that elsewhere I have called "rhizo-ethology" (Probyn 2004a). It is indebted to Gilles Deleuze's understanding of ethology and the use of the rhizome as a way of figuring relations. Through ethnographies I recount the rich materiality of fish and humans. I use ethology as a way of figuring their relations. As will be familiar to many, Deleuze and Guattari (2004 [1980]) distinguish the rhizome from arboreal thought, the latter a rude way of framing cause and effect. Following rhizomes requires questioning how things (ideas, histories, environments, biological entities) connect, and inquiring after the work they do in certain milieus. This approach instills modesty and enacts an ethical sensibility because you do not know where, why, or how the shoots of a rhizome will next erupt. This is clearly stated in Deleuze's (1992) essay "Ethology: Spinoza and Us." The phrase is meant to evoke "us in the middle of Spinoza" (Deleuze 1992, 625). For Deleuze, this entails "the laying out of a *common plane of immanence* on which all bodies, all minds and all individuals are situated" (625).

The image that this brings to mind is of the ocean laid out in such a way that all her relations are clearly seen in connection to each other. For instance, in my opening evocation of Sydney Harbour, ideally I'd like to be able to figure all the dimensions and scales at work, at once temporally as well as spatially: the boats, the fish beneath, the people alongside, the reasons why the water is polluted, the toxic fish, the flourishing fish, the fish that are eaten, the humans that ingest them, the maladies that follow in both fish and human communities. This is a multidimensional figuring that I try to grasp in its complexity. The ocean is a Spinozan body par excellence. It is composed of an infinite number of particles: "It is the relations of motion and rest, of speeds and slowness between particles that define a body, the individuality of a body." And it is "a body [that] affects other bodies, or is affected by other bodies" (Deleuze 1992, 625). Deleuze writes that these two propositions are simple; one is kinetic and the other dynamic. But it is in the middle that "things are much more complicated" (626).

Think of being in the middle of the ocean, of always being in its middle. Most humans know the ocean from its edges, standing on the liminal shore looking out. But from the middle we may envision the "complex relation between differential velocities, between deceleration and acceleration of particles" (Deleuze 1992, 626). At each and every turn things are being affected and are affecting. To return to oysters again, they are very much bodies that affect and are affected by other bodies: they are in and with the estuary, the loch, tides and human tastes, natural and man-made histories. Oysters tinker with nature, itself understood as an assemblage, a body. As a keystone species, oysters do work far beyond their size. An oyster filters up to fifty gallons of water a day, cleaning the water of other bodies—such as the nitrogen excreted through the more-than-human practices of agriculture. As Stefan Helmreich writes, "Human biocultural practices flow into the putatively natural zone of the ocean, scrambling nature and culture, life forms and forms of life" (2009, 13).

Other sites tell of different forms of human-fish-ocean relations. Bluefin tuna encapsulate another way of navigating stories of globalization. This warm-blooded body, the fastest fish, easily travels the globe and then gets caught in webs of the meaning making of human greed. Bluefin tuna becomes a marker of cultural and economic capital, as it also changes people's tastes around the world. We encounter the men who have homed in on them, following the fish from Croatia and Italy to Australia.

Ethology studies "the compositions of relations or capacities among different things" (Deleuze 1992, 628). This is central to *Eating the Ocean*. As I've argued before, we eat and are eaten (Probyn 2000). There is no privileging the inside or the outside of any individual body. If one eats bluefin tuna, one eats at the top of the trophic system, ingesting the heavy metals the tuna has eaten across its history. Human eaters get a taste of what we have wreaked. We eat oil slicks, and the chemicals used to disperse them eat into our flesh. Fish eat the microplastics used in daily skin care; humans eat the fish and the microplastics; and fish and human bodies intermingle. And of course that "we" gets eaten up too, differentiated, fragmented, and fractured. Moira Gatens argues that we need to add historical depth to Deleuze's ethology. What is the genealogy of how bodies come to be figured in certain ways? As she writes, "Given that on the ethological view nature, bodies, and materiality itself, is active, dynamic, and has a history, then past compositions will affect the present and future possibilities of what we may become" (Gatens 1996, 12). This raises the question of how gendered and classed bodies have been erased not only in the histories of fishing but also in the current theories of the more-than-human.

In their articulation of what they call "a wet ontology," Kimberly Peters and Phillip Steinberg ask, "How can one write about so 'slippery' a subject[,] . . . how can one write about the ocean as something to think not only *with* but *from* without reducing it to a metaphor?" (2015). Fish for me are not, and cannot be, metaphorical. Fish are the very principle of relatedness that holds this book together. But they are, after all, often below the surface. This is why I insist on an embodied and dialogic ethnography that is attuned to listening to stories and relaying them, to trying to capture affective spaces through various forms of description, and to reaching for the depth of history that informs tacit knowledge embodied in individuals' ways of being and ways of recounting.

This informs a certain form of writing and an embodied engagement. Focusing on the relatedness of fish and humans also requires relating and telling stories, often from oblique angles. This will no doubt infuriate some readers. But this mode of research has reason; its rationale is in what it can or cannot do. My embodied, dialogic method also has form. While the very nomenclature of the more-than-human would seemingly demand an emplaced way of relating, its genealogy is often forgotten. In the rush to champion new nonrepresentational methodologies, potentially much gets thrown out. For instance, Nigel Thrift (2008) defines the nonrepresentational as anti-autobiographical. This is an implicit disregard and miscomprehension of much of feminist writing inspired by the lived fabric of the everyday, too often derided as the realm of "the personal." Against this, Hayden Lorimer argues for "work that seeks better to cope with our self-evidently more-than-human, more-than-textual, multisensual worlds"

Fig. I.2. Fish–women entanglement. Illustration by Morgan Richards.

(2005, 83), which he characterizes as framed by "a cultural-feminist programme that has nudged the more-than-representational debate out of a predominantly white, western orbit" (89).

Gendering and Queering More-Than-Human Fish

Throughout this book, my own genealogy as a queer feminist tugs at me. (Figure I.2 captures some of that swimming with and against in a jumbled-up sea that is so very queer.) At different iterations of this project I have tried to shrug off that history. In Australia, as elsewhere, being a university researcher involves applying for major government funds. This requires being sensible, making sense to committees composed of bodies disciplined to conceive of the world in rather narrow proprietorial terms. Worrying that scientists would not understand why a researcher with my background in queer feminist theory would be interested in fish, I tried to hide from myself. I straightened myself. I took on the label of "social scientist" as if it came with a white coat that would allow me entry into quarters where other scientists converse. If needs be, I can walk the talk. But I couldn't write this book trying to pass as something other than what my intellectual history has made me. And trying to pretend I could was doing my head in. So my queer feminist self took over. To slightly reframe Kathryn Yusoff's words, while "there is something lonely, yet necessary in this act of making [feminist] relations . . . it allows us to get over ourselves and seek out what is truly strange and wonderful in the cohabitation of worlds we will never be at home in" (2013, 225). Doing the research that fuels this book has, if anything, made me a more adamant queer feminist fish.

Not quite belonging in the circles in which I swim has also made me appreciate the passions and deep investment of marine researchers in their various specializations. While they are sometimes surprised that I am passionate about their work, they are very aware that engaging with the ocean requires understanding humans. They even get that gender is an important part of the puzzle. Equally, I've found the work of many feminist social scientists crucial to my understanding. Women—as researchers, fishwives, fishers, oyster growers, NGO workers—are deeply enfolded in fisheries, but as we know, if they are not counted, they do not count. Barbara Neis is blunt: "Gender relations permeate fisheries at every level." And women's ecological knowledge has been mediated "through their relationship with men—fisher*men*, husbands and sons, male-dominated governments, and male-dominated science and industry" (Neis 2005, 7). Count the fish until they are gone; don't ask the women what they know. Don't count on women's experiences. But around the world, women's voices are beginning to be heard. Groups like Genderfish conduct research and provide training to women in the Global South. In Australia I've joined with groups like the Women's Industry Network Seafood Community, which conducts workshops on different aspects of the industry and especially how to promote women's leadership. These groups face similar issues such as suicide rates among fishers, routine sexism, sexual violence, and slavery in the global fisheries.

For Helmreich "the ocean is strange" (2009, ix). In his *Alien Ocean*, Helmreich argues for what he calls "athwart theory" (23). This he describes as "an empirical itinerary of associations and relations, a travelogue which, to draw on the nautical meaning of *athwart*, moves sidewise, tracing the contingent, drifting and bobbing, real-time, and often unexpected connections of which social action is constituted, which mixes up things and their descriptions" (23).

To Helmreich's use of the nautical sense of "athwart," I add Eve Sedgwick's understanding that "the word 'queer' itself means *across*—it comes from the Indo-European root *-twerkw*, which also yields the German *quer* (transverse), Latin *torquere* (to twist), English *athwart*" (1993a, 12). To work athwart is for Sedgwick to be within the spheres of "continuing moment, movement, motive—recurrent, eddying, *troublant*." Troubling eddies indeed.

"Athwart" closely describes this book. Here we travel sideways across different bodies of disciplinary knowledge, across the scales of the intimate and the public, of poems, literature, government reports. And in the fieldwork we move from the bottom of Australia to the top of Scotland, follow sardines from California to the anchovies of Peru that end in the fish farms of Tasmania and Scotland. Bluefin tuna take us from the east coast of North America to South Australia and then to Tsukiji, Tokyo's fish market. Bodies meet bodies continually. We encounter "herring lassies covered in fish guts . . . [so] 'bespattered with blood and the entrails and scales of fish as to cause them to resemble animals of the ichthyological kingdom'" (Charles Richard Weld quoted in Nadel-Klein 2003, 81–83), and flirt with elderly tuna barons and retired skippers.

As I said at the outset of this introduction, the questions I ask and the manner in which I relate different dimensions seek to foreground the necessary complexity with which we must approach fish, human, and oceanic relations. There is no doubt about it; viewed from any number of angles, the situation is not good. And yet as I argue, we need to learn to care in new ways about this very old relation of humans, oceans, and fish. I hope that a certain exuberance engages my readers. The necessary complexity of my subject matter meets the necessity of inhabiting deeply the otherworldly spaces of fish-human relations. My ethnographic and ethological methods take us into the thickness of the oceanic. Immersed in the storied relation of fish-human encounters, there is no possibility of returning to the safe shores of simplified food politics. Caught up in the flotsam and jetsam of the tales in this book, the stakes remain stark: How can we eat better with the ocean? What, in practice, does it mean to be athwart the ocean and her more-than-human dependent inhabitants? Several elements distinguish a marine-based ethics of food from the dominant terrestrial food politics. One is, as I've flagged, geographical and geopolitical—the complexity of fish-human-food entanglement draws us into other worlds: the lives of men and women and fish in different parts of the world. For instance, the fish sold in Billingsgate Market in London come from all over the world and

Fig. I.3. Billingsgate Fish Market, London. Photograph by author.

are sold by buyers whose forebears were subjugated as part of the British colonial enterprise (figure 1.3). Some would be deemed good fish, such as the tilapia farmed in the Nile (although not in Australia, where it is a feral fish endangering other species), or the mackerel from India, or indeed the cheap sardines. In Billingsgate they are all called exotic fish, because they are destined for migrant tables.

If, as Kathryn Yusoff argues, "the *making sensible* of biotic subjects is a basic tenet of conservation practices" (2013, 209), fish-human entanglements refuse to be made sensible. That fish refuse to settle into a neat taxonomic order, to cuddle up to us, is important for an ethics of food that departs from human anthropomorphic desires. The other element that is important, and to which I turn in chapter 1, is the seeming indifference of the ocean toward human life and lives. Being caught in a rip current or being at sea in a storm reminds us of the sheer power of the ocean that can leave us speechless. In Yusoff's terms, "the recognition of that which cannot, and will not, be brought to sense requires a response, then, that is not configured through a mode of auto-affection, but through a mode of relating that is indifferent to 'us' and holds fast to that indifference" (2013, 209).

Combining an appreciation for that indifference with a desire to learn, to relate the stories of others, is to be athwart the ocean and her dependents—human and fish alike. It's not a particularly comfortable position, but it brings with it a sense of awe, of wonder, and I hope the desire to learn more about our fish-human entanglement.

1

An Oceanic Habitus

Alone, alone, all, all alone
Alone on a wide wide sea!
—Coleridge, "The Rime of the Ancient Mariner," 1834

It is so vast, the ocean. In this first chapter, I bring together many ideas from very different sources to navigate what a kayaker would know as clapotis—where incoming waves meet with rebounding waves. It's an experience that can shake you up. Sometimes it's not bad to be all at sea—disoriented, we may begin to see new directions. Here I begin to parse the question of fish and human relations. Throughout this book I make an argument—or several—against simplistic options, solutions, reactions . . . the pairing of the familiar us versus them, even as the actors interchange.

In public and academic debate, the cultural politics of food continues to be a powerful if conflicting site where forms of state policy and economic, cultural, and affective investment all compete. Individual, regional, and global concerns are all at play within this fraught sphere. It is certainly not a new area, and questions about how we are to feed humanity, and with what, have been ongoing for decades if not millennia. But increasingly it strikes me that the framing of these deep-seated questions strips them of some necessary complexity. In marine science, researchers now talk of the "simplified sea" (Howarth et al. 2014). This describes the problems caused by "fishing down" or "fishing through the food web." This is producing a vastly changed sea, a simplified sea where biodiversity has been stripped away. Once the big predators have been wiped out, "ecosystems become dominated by a handful of species such as prawns, lobster, macroalgae and jellyfish that used to form the diet of, or were outcompeted by, larger fish" (Howarth et al. 2014, 691). Bizarrely enough, there is a temporary upside to this pillage of oceanic biodiversity. In places like Maine and Nova Scotia, the west of Scotland, and Tasmania, some fishers are happy. As the big fish are depleted at the top of the food chain, the smaller ones down the web get their turn. Lobster catches have exploded over the last several years, as too have scallops and prawns. And as one could guess, the value of these invertebrate catches is many times that of the previous resource. It's a bizarre state of affairs.

I want to use the phrase "a simplified sea" to frame some of the initial thoughts that flow through this chapter. First, let me explain more precisely what marine scientists mean by the term. It has to do centrally with oxygen, which is not a substance most of us think about in relation to the ocean. As the big carnivorous fish become ever more scarce, the phytoplankton flourish—in a bloom these tiny floating plant organisms can increase a thousandfold. But they are short lived, and when they die off, bacteria consume them. This eats up more oxygen. Organisms slightly higher up the chain, such as mussels, can't get enough oxygen from the water. They then die off, which produces yet more decaying material and further depletes the oxygen in the water, which has now become thoroughly hypoxic. This provides just the niche that jellyfish love, and they can rapidly take over huge areas. As carnivorous organisms, they eat the eggs of any remaining fish (Howarth et al. 2014, 696).

This process is also called trophic cascading, where like a knife through butter one shock causes other shocks down through the complex webs of life in the sea. The result is dead zones in the oceans around the world where nothing can live. Simplified systems just don't have the resources to survive man-made or natural shocks. I am going to take this term to describe how the ocean and her dependents are being simplified across different registers of cultural representation.

The stories I collate and tell across this book, in their different ways, speak of a necessary complexity. The complex interactions of the highly diverse systems housed within even more complex ecosystems are for me a cue to up the ante against the simplified answers that are routinely trotted out by well-meaning organizations. I'll get to these in more detail later, but if the answer to the problems of fish and the oceans is no, then the question is seriously simplified. I remain at heart a Deleuzian, which is to say that I follow multiple entryways into the entanglement of humans and nonhumans, into our vexed encounters within different ecosystems. The point for me is to diagram, to model, to figure, and to story their interactions in ways that may proliferate different angles, optics, and perhaps understandings.

Turning away from the terrestrial and to the ocean compels an alternative way of thinking about the enmeshed human and nonhuman ecosystems. Perhaps because most humans can't live in water, until recently human concern has been directed to the terrestrial engagements of humans and animals. Food politics, for instance, has been overwhelmingly focused on terrestrial animal protein. Is this simply because it's easier to care about a cow than a lobster? Classic animal rights texts such as Peter Singer's *Animal Liberation* hesitated about where to draw the line. In the 1990 edition, Singer recounts, "With creatures like oysters, doubts about a capacity for pain are considerable; and in the first edition of this book I suggested that somewhere between shrimp and an oyster seems as good a place to draw the line as any" (174). For many, pescetarianism (fish-eating vegetarianism) makes sense: For some, it's about not eating an organism that can cry out to human ears; for others, the driving question is how to feed the planet equitably and efficiently (and of course, land-based animals are not efficient as a means of producing protein). It sounds mercenary to talk in terms of animals as efficient or inefficient producers of protein for humans, but a looming population of nine billion requires critical thought. And even though aquaculture is in its infancy compared to agriculture, the feed conversion ratio of fish is in many instances lower than that of land-based animals—I return to this issue in chapter 5. Of course, many people choose not to eat meat because of the closeness of mammal eating mammal. Concern for some species over others may be simply because of

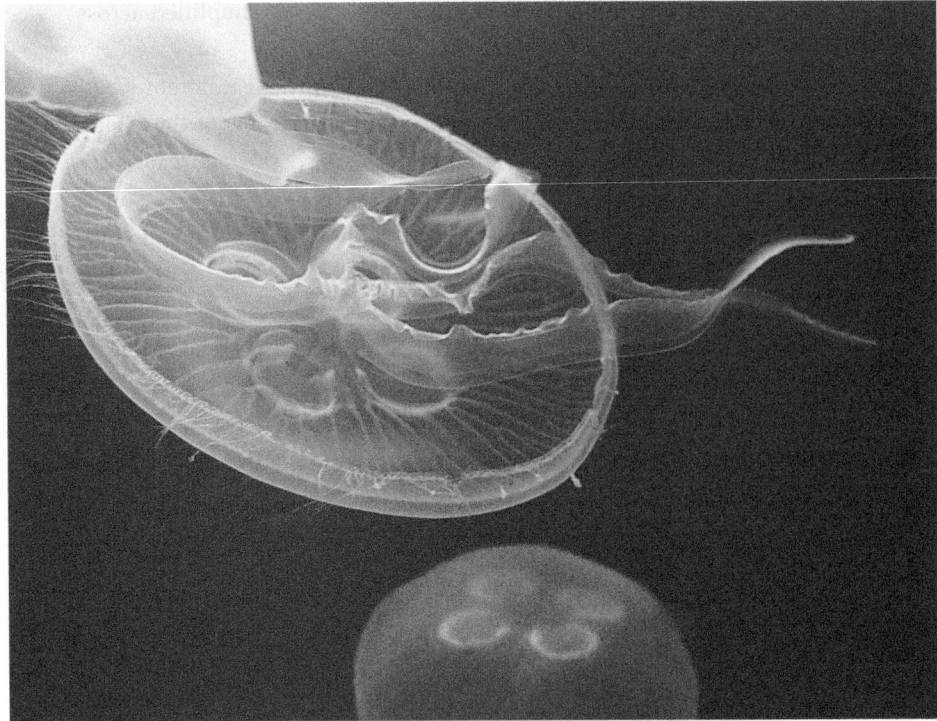

Fig. 1.1. Jellyfish at the Aquarium de la Porte Dorée, Paris. Photograph by author.

their good looks and good luck to be anthropomorphically cute. It's hard, though not impossible, to cuddle a fish.[1] Anthropomorphic projections can belittle animals, but they can also afford an imagined connection between and among species. In a future life, I want to be a bluefin tuna or a jellyfish. How we imagine and engage with the complex issues of human-fish relations may effect which animal is still extant.

Christopher Bear and Sally Eden argue that many of the new animal geographies exclude fish because of the "alien" spaces they inhabit: "Water environments [contrast] with the 'airy' spaces that we humans inhabit" (2011, 337). It's an environment where humans often get horribly seasick. I certainly do. But this unsettling queasiness can be productive. Gisli Pálsson, an Icelandic anthropologist of fish, understands seasickness as a kind of culture shock, which is produced out of and indicates a lack of "emotional and physical manifestations of mastery and enskilment" (1994, 905). In a different context, and with different effects, sailors in the tropics were said to suffer calenture, fevers that caused them to want to throw themselves from the boat into what they imagined as cool billowing green pastures.

In moving out of the realm of arbitrary hierarchies of what is good or not good to eat, can we engender a more wide-ranging cultural politics of sustainable food production and consumption? As marine biologist Carlos Duarte asks, "Will the oceans help feed humanity?" (Duarte et al. 2009). To which it needs to be added, is it too late? Is the ocean broken (Ray 2013)? In early 2013, Ivan Macfadyen sailed his yacht from Australia to California two years after the Japanese earthquake, the tsunami, and the disaster of the Fukushima nuclear reactor. Macfadyen describes how "only silence and desolation surrounded his boat, Funnel Web, as it sped across the surface of a haunted ocean" (Ray 2013). Without simplistic extremes like T. H. Huxley's (1882) "inexhaustible" ocean fisheries or the image of the ocean on its last legs, how might we formulate a cultural politics that can encompass the vast challenge of sustaining fish-human-ocean relations? There is little consensus about what action to take, or which aspect to focus on.

To illustrate this, I want to briefly analyze several recent representations that foreground different understandings about the current state of fish and fishing. What forms of affect may or may not be mobilized within a cultural politics of oceanic connection? What types of action might be envisioned and hoped for? What oceanic complexities are silenced?

The End of the Line: Imagine a World without Fish (Murray 2009) is a gripping documentary about the interconnections that imperil fisheries and oceans worldwide. Based on the British journalist Charles Clover's book of the same name, it is narrated by Ted Danson and has the high production values more common to an Attenborough blue chip nature documentary (Richards 2013) than the genre of animal rights activist film. It had a budget of over a million GBP, cobbled together from the UK Channel 4 Britdoc Foundation and numerous smaller agencies. It is a highly researched and passionately told tale of what is happening in the oceans. It features renowned fish scientists such as the University of British Columbia's Daniel Pauly. Pauly is a charismatic scholar. He looks directly into the camera and states in his lightly French-accented voice, "All the fish are gone. Where are they? We have eaten them." I get goose bumps. Of the documentaries I have viewed over the course of this project, *The End of the Line* is the most powerful. The production team brought on board organizations as different as Waitrose and Greenpeace, and apparently the documentary caused several businesses to change their sourcing practices. A YouGov survey of two thousand people commissioned by Waitrose in 2009 found that "when made aware of the facts, 70% of people are more likely to make sustainable choices. But 78% admit that they currently don't attempt to buy sustainable seafood at all" (Channel 4 Britdoc Foundation 2009).

Sadly, this is all too common. What people say they want to do (or even do want to do) is often at odds with what they actually do. But *The End of the Line* did create ripples. It spawned several social media campaigns, such as Fish Fight, an initiative of the British celebrity chef Hugh Fearnley-Whittingstall. Fish Fight had one goal when it started in 2010, and that was to force the European Union to change its regulations to ban the practice of throwing back discarded fish. Perhaps because it was focused on one objective, and it used social media to mobilize large numbers of people in the United Kingdom to pester their members of European Parliament, it worked. The EU changed its regulations.

The End of the Line and Fish Fight may also have led to a television documentary narrated and promoted by Sir Richard Branson and Virgin, called *Mission: Save the Ocean* (Nat Geo Wild 2013). Much of the documentary is shot in thriller mode and combines glorious shots of Branson's Necker Island with edgy noir scenes. The stars are the big three ocean and fish sustainability programs: Greenpeace, the Marine Stewardship Council (MSC), and the World Wildlife Foundation (WWF). It seems that Sir Richard really does care about the ocean; he is an Ocean Elder—one of a rather diverse collection of high-profile people concerned with ocean awareness (including Ted Turner, Prince Albert II of Monaco, Jackson Browne, and Sylvia Earle). Branson's message is squarely on the power of the entrepreneur and of the consumer: "I'm an entrepreneur and entrepreneurship is about new ideas and making a difference. We can all make a difference. Your choices in everyday life can actually make things better for the ocean" (Nat Geo Wild 2013).[2]

My next example is a 2010 Dutch documentary, *Sea the Truth* (Everaert and Zwanikken 2010), funded by the Netherlands' Party for Animals and the Nicolaas G. Pierson Foundation. It followed from a previous documentary called *Meat the Truth* (Soeters and Zwanikken 2008), which encouraged people to give up eating meat for a day, a week, or a lifetime. *Sea the Truth* is narrated by the leader of the party, Marianne Thieme, and centers on Dos Winkel, an activist, scuba diver, and photographer. While his art is spectacular, much of the documentary focuses on his talking head. The main issues are overfishing for human consumption and the practice of using fish to feed animals and other fish. The process of fish reduction is represented in cartoon form, and it does capture the lunacy of turning fish into fish meal, and the complicated issues of the production and consumption of fish oil capsules (to which I return in chapter 5). Echoing Huxley, Thieme speaks of how the ocean was "an inexhaustible horn of plenty for humans." In her conclusion, there is but one player to be blamed: "The fishing industry is responsible for the disappearance of species and the destruction of valuable ecosystems and that is also true for so-called sustainable fisheries. The entire industry runs off billions of government subsidies. Every citizen is paying for the destruction of the seas and oceans."

Thieme frames "the fishing industry" as one undifferentiated, monolithic entity. Many people assume this without realizing how complex and distinct each form of commercial fishing is. The only thing that holds them together is the ocean, and even then the differences between inshore fishing and deep-sea fishing are immense. I have attended many fishing association congresses, and one thing is clear: The industry is not one. It is one of the most fragmented collections of commercial endeavors. On the one hand there is the myth (and at times reality) of the rugged individualist fisherman loath to tell of his secret fishing spots, and on the other there are the very different practices included under the rubric of commercial fishing. Some fisheries are certified as sustainable; some are not certified but practice responsible fishing; some are criminal operations using indentured labor to scrape the seafloor bare. But there is no one industry to finger. Ignoring this complexity, Thieme calls on the consumer to resolve the situation: "Our forks are mighty weapons. Use them for a sustainable future." This catchy line erases from view the complexity of the chains that lead from production to consumption.

The final example I will mention is the Australian-based documentary *Drawing the Line* (Blyth and Gloor 2013). The title is a direct if unspoken riposte to Clover's book and documentary. The tagline is "What if you lost everything you loved because someone else wanted to protect it?" "Drawing the line" also refers to the lines being drawn in the sea for marine protected areas (MPAS). Australia is, of course, an island nation. Currently planned MPAS will constitute some 30 percent or more of the world's total MPAS. The documentary is unabashedly partisan—funded by and starring several Australian fishers. To back their line, the documentary includes interviews with some of Australia's top fisheries scientists, such as Colin Buxton and Caleb Gardner from the University of Tasmania's Institute of Marine and Antarctic Studies.

Marine parks are particularly complicated in Australia, as the commonwealth and the states and territories have different areas of jurisdiction and management control. Marine reserves in commonwealth waters start three nautical miles (5.5 kilometers) from shore (Department of the Environment 2014). This means that the all-important recreational fishing sector is mostly spared by commonwealth measures. This sector is very politically savvy, with campaign slogans such as "I fish, I vote." They had the former Australian prime minister, Tony Abbott, completely on board, appearing in countless photo ops looking Putinesque: macho political leader complete with fishing gear.

In the public realm, there is little concern for commercial fishers. Such is the fishing industry's worry about the public's perception of the fishing industry that many fishers now call themselves "professional" rather than "commercial" fishers. It's an interesting divide: Fishing for a living is portrayed as pillaging the seas for immense profits, whereas recreational fishing is seen as a benign activity despite its practitioners catching millions of tons of different species, including valuable ones such as lobster and abalone. *Drawing the Line* relies on several Australian fishing families to convey how deeply those who live on the ocean feel about "the industry that is part of the family." The documentary's message is that Australian fisheries are the most highly regulated and managed in the world, with which many fisheries scientists would concur. By and large Australian fishers have come to grips with the quota system; however, the government's turn to MPAS as the preferred measure for regulation is deeply upsetting to the fishers as it effectively closes off large segments of their previous fishing grounds.[3] The fishers portrayed in the documentary are like many in Australia, multigenerational family businesses who regard themselves as long-term custodians of the resource by which they live.

Across all the documentaries, one cause is to blame. For *Drawing the Line* it's the government; for the others it is fishers and the fishing industry. But does it need to come down to a choice between the fish or the oceans or the fishers? Formulating a cultural politics to sustain fish, fishers, and oceans is undoubtedly complex and downright hard. However, these documentaries smooth out complexity to yield simple take-home messages. That the messages are all different belies the complexity we would rather not face.

And each differently simplifies the sea. The message and the imagery of the most affective as well as effective one, *The End of the Line*, draws us into the world of the global violence of illegal trawlers. It is the most effective because it is the least simplistic. However, at the end of the day the message is simple: stop eating fish. The facile politics of *Sea the Truth* promotes the idea that we can resolve everything by "voting with your fork." Here the political, cultural, and economic implications of choice are effaced. There is no downside in sight: Save the fish; save the oceans from humankind. But of course in so doing we say no to millions of fishers. We may be also saying no to more effectively figuring the complexities of ocean, fish, and human relations. To return to those depressing figures about what people say rather than what they do, part of the problem may be how these documentaries take what is a complex yet fascinating set of questions and represent the issues of choice. But what we see is that stripping the more-than-human ocean of its intricacy doesn't seem to make people care more. Against this simplification of the sea I think we need to ramp up reasons why the sustainability of fish and human culture matters, and for whom.

The Problem with Sustainability

To be blunt, science is hard pressed to make us care. Surely that's the task of cultural theorists and social scientists. But if the fishery industry is fragmented, so too is cultural and social theory—fractured and sometimes fractious. Of late some literary theorists are challenging the very idea of sustainability. For instance, Steve Mentz argues that "the era of sustainability is over" (2012, 586). It's not that Mentz thinks that all is well. Rather, he critiques what he sees as a dominant "pastoral nostalgia" strain within the sustainability discourse (587). Mentz argues that "if we turn from green pastures to blue oceans, we find an already present, partially explored environment for post sustainability thinking" (586). Mentz wants to mine the ocean as radical alterity for whatever might come after sustainability. His manifesto for "blue cultural studies" distinguishes between the safe pastoral and the wild ocean: "For many early modern writers, the land is orderly and human; the sea chaotic and divine" (Mentz 2009b, 1001). He cites Plato's caution in the *Laws*: "For the sea, although agreeable, is a dangerous companion, and a highway of strange morals and manners as well as of commerce" (Mentz 2009b, 998). Plato's words capture much of what still happens on the wild open seas—slave labor, piracy, deep-sea mining, and shipping. For Mentz, the ocean is a seductive place in which and on which to reflect: from the sheer physical incompatibility of human beings in salt water—which Mentz sees as "the richest vision of a nonhuman nonsustainable ecology" (2012, 587)—to his oceanic readings of Shakespeare (Mentz 2009a) and his "swimmer's poetics" that celebrate "giving oneself over to the alien element" (Mentz 2012, 589). As a literary critic, Mentz brings the literary to bear on "encounters between human experiences and disorderly ecologies." This is, then, to attempt a poetics of embodiment: "Through these encounters, we learn what living in a post sustainable world feels like on our bodies, as well as how to devise textual structures to make sense of disorder" (Mentz 2012, 588).

Stacy Alaimo, writing from the perspective of feminist posthuman studies, also takes issue with the sustainable, asking, "What is it one seeks to sustain, and for whom?" (2012b, 561). Her remarks are confronting for my project: "The promotion of, say, sustainable seafood assumes that there are marine creatures one can consume without threatening their continued existence or becoming harmed oneself" (Alaimo 2012b, 562). There is, of course, no denying that eating fish, even low on the trophic scale, is still consuming and terminating their individual existence, although not necessarily the entire species. Alaimo is concerned with some bodies but not others. For instance, she takes up the tale of Hardy Jones, an ocean conservationist who has devoted his life to saving dolphins and whales (BlueVoice 2008). Several years ago, Jones discovered that he was a victim of mercury poisoning, just like the dolphins and whales with whom he spent his life. He, like them, was a top predator, and his body became a receptacle for mercury and toxins mainly through eating tuna and swordfish. Mercury gets into the oceans primarily from coal-fired power plants, and from mining and agricultural runoff into the ocean. It is ingested by algae that process it as methylmercury, which then is bioaccumulated and intensified throughout the food chain, rendering long-living large fish-eating fish particularly vulnerable. Several populations are especially vulnerable. For instance, the Inuit are particularly at risk because of their diet high in fish but also because methylmercury is even more highly concentrated in Arctic waters. In 2003 Jones was diagnosed with an unusual form of blood cancer. The organochlorines from pesticides were to blame for both his cancer and the increasing number of cases of cancer in marine animals.

For Alaimo, "sustainable" has come to dominate our understandings of the human and nonhuman environment. She asks, "How is it that environmentalism as a social movement became so smoothly co-opted and institutionalized as sustainability?" (Alaimo 2012b, 559). This is an important point. From being a marginal concern, corporate sustainability schemes and marine certification schemes are now big (not-for-profit) business. For example, the MSC charges fisheries for use of its eco logo once they have gone through the third-party process of being ascertained as sustainable. The council was a part of the Unilever-funded WWF but became independent in 1999. It is increasingly coming under attack for giving out the certification too easily (Smith 2011). Unbeknownst to most consumers, it is the fishery that funds the costs of a lengthy and expensive process of being certified as sustainable. While the large fisheries can swallow these costs, the smaller and often more inherently sustainable fisheries are often hit hard by the price of getting the eco logo—that more often than not goes unnoticed by potential consumers.[4]

In Alaimo's framing, there is sustainability as big business versus the environment as social movement. This division effectively eliminates from the frame the multiple actors that need to be brought together if sustainability is to be more than just a corporate buzzword. Elsewhere Alaimo specifies the players in her vision of transcorporeality: "Activists, as well as everyday practitioners of environmental health, environmental justice, and climate change movements, work to reveal and reshape the flows of material agencies across regions, environments, animal bodies, and human bodies" (2012a, 476–77). It seems only a select few are granted the presumption of a more-than-human transcorporeality. But what of those who spend their lives in and on the sea? In framing an inclusive modality of caring for the more-than-human, surely we can't afford to ignore those placed outside of Alaimo's frame: the fishers, regulators, and indeed even consumers of fish?

In a special issue of *American Literature* devoted to ecological matters, another framing of a disciplinary and political divide emerges. In their preface to the issue, M. Allewaert and Michael Ziser argue, "Our contributors' responses to this ['ecologico-economic'] state of affairs takes us away from the place-based, policy-focused, and phenomenological preoccupations of older forms of ecocriticism towards an engagement with murkier aspects of our condition: the historical imagination as an alternative resource for a world facing shortages in the feedstocks of modernity; the significance of new identities and communities—often involuntary—that arise from environmental crisis; and the complex new forms of affect that accompany the re-categorisation of the planet" (2012, 235). Here "the historical imagination" is opposed to "place-based, policy-focused, and phenomenological preoccupations"; murkiness trumps clarity of previous approaches; and new forms of affect oppose older identity-based forms of action and politics.

My argument with these approaches is that they are fundamentally divisive and exclusive—the new versus the old, and some players rather than others. Surely the problem is an expansive one, large enough for all. They also seem to exist within a bubble untouched by the context of global concerns such as malnutrition and the livelihoods of often small artisanal fishers and their families. To return to my argument for necessary complexity, the discarding of certain actors—be they fisheries scientists or bad consumers—produces a simplistic cultural and social ecosystem. This contributes to a sort of intellectual hypoxia, which strips our capacities to more widely imagine what sustainability could be.

Affective Habitus

How do we come to care for complex entanglements? I'd argue that it is through detail, through relating stories and experiences, through paying attention to the intersections of human and fish lives. Consider figure 1.2. It's a black-and-white photograph of Footdee, pronounced Fittee, the old fishing village of Aberdeen. As you can see, it's a fairly bleak place, and even in real life the predominant tones are gray. Some sort of settlement has stood there since the fourteenth century. The squat houses built of granite have stood their ground against wild weather and the failing fisheries. While some are gentrified, one house in particular caught my eye for its enduring remnants of life long lived by the sea—the washing blown in the sea winds that never quite dries, the garden carefully tended. It will be a while before spring reaches the chilly northeast, and even then the daffodils will be battered by the gales. Coming through the harbor is not a fishing boat but a "mother ship" to take supplies out to the oil rigs. This is a story of fish long gone, replaced by the unnatural mining of the seas. The grandsons of the fishers, if they are lucky, get long shifts on the rigs. The women make do as they can. Young and old drift. It's a story of local multigenerational unemployment and obscene wealth for the multinational oil companies. It's said that the oil is going too, taking the money to another part of the world.

Fig. 1.2. Mother ship going out at Footdee Harbour, Aberdeen, Scotland. Photograph by author.

How we care or don't is very much a question of habitus.[5] While Bourdieu's conception of the habitus has its limitations, it nonetheless continues to be useful in laying down the conceptual bases of an individuated socioecological sphere. He, of course, frames it as the embodiment of the social and cultural, or more precisely thus: "A product of history produces individual and collective practices—more history—in accordance with the schemes generated by history. . . . [The habitus] ensures the active presence of past experiences, which, deposited in each organism in the form of schemes of perception, thought and action, tend to guarantee the 'correctness' of practices . . . more reliably than all formal rules and explicit norms" (Bourdieu 1990, 54).

That our habitus inflects or orientates what and how we are is at one level the stuff of common sense. From the learned acquisition of manners, of how to handle knives and forks or chopsticks or with which hand to eat, to the acquired tastes we learn to like (or learn that we shouldn't like), these are practices that, as Bourdieu writes, are "something *that one is*" (1990, 73). Bourdieu's longtime collaborator Loïc Wacquant draws attention to habitus as "the way society becomes deposited in persons in the form of lasting dispositions, or trained capacities and structured propensities to think, feel and act in determinant ways, which then guide them" (2004, 316). More recently Wacquant has returned to defend his late colleague against the accusations that "habitus is a 'black box' that muddles the analysis of social conduct, erases history, and freezes practice in the endless replication of structure" (2014, 193). It is true that at times Bourdieu's concepts seem static, but Bourdieu also argued that "habitus change[s] constantly as a function of new experiences" (Wacquant 2014, 195). Wacquant revives Bourdieusian theory to argue that it can and must be "an approach that takes seriously the embarrassing fact that social agents are motile, sensuous, and suffering creatures of the flesh, blood, nerves and sinews doomed to death" (2014, 198).

I wonder whether habitus can help us understand how bodies open to other bodies, human and nonhuman? Certainly an attention to dispositions is crucial to my argument—these are bodily orientations, ingrained ways of embodying the world, which while they may change are truly our history forged in flesh, taste, and memory. As I have argued elsewhere, we need to foreground the role of emotion to frame an affective habitus (Probyn 2004a, 2005a). This is to turn Bourdieu's attention to the bodily imbrication of societal structures and extend it to the forms of affect and emotion that the body registers in the incorporation of the social. As I've put it elsewhere, the body eats into the social as our bodies are simultaneously eaten by practices that are the instantiation of class, gender, ethnicity, and so forth (Probyn 2000). While of course Bourdieu's work continually elaborated on the body as "hinge" between the objective and the subjective, I want to give this more bite, so to speak. How and why we care about things, people, and places is a continual process whereby "caring" can hurt or reassure or be joyous, as new knowledge, new ideas, different practices intersect with the primary habitus. This could be called a coming to care. In Beverley Skeggs's work, for instance, the yearning for "respectability" deeply informs what the abstract "working class" may mean for some women. For Skeggs, "practices such as respectability, assumed to be middle-class, are significantly reworked and revalued when lived by the working class: a complete ethical re-evaluation" (2004, 88). In this argument, Skeggs reworks Bourdieu's habitus to allow for change, incorporation, and repudiation. In terms of what I am calling an affective habitus, this allows a window of opportunity to consider how people might or might not incorporate forms of care as something transactional, as something that is ingested and digested—a reworking and an opening out of the self. This thematic is woven throughout this book—that the many more-than-human actors are "feeling and desiring animals who know the word *by body in practice*" (Wacquant 2014, 197).

To return to my question, how do we embody a care for the sea and its dependents? Some of us humans are seemingly out of our element when it comes to the seas and oceans, although many argue that humans evolved from the sea. But we—the nonseafarers—are by and large lost in the waters of what we call Earth, which is of course three-quarters seawater. Stefan Helmreich's work is instructive here, arguing as he does that while seawater has long been implicit in cultural theories and in fact underlies many of the now-overused metaphors of globalization (especially "flows"), we need to bring to the surface the materiality of the oceanic. One of Helmreich's points is particularly salient. He proposes the idea of "athwart theory," which he takes from its nautical context—as in across a ship's course or across its decks (Helmreich 2011, 134). I cannot hear the term without thinking of Eve K. Sedgwick's immemorial definition of queer. As she wrote in *Tensions*, "The word 'queer' itself means *across*—it comes from the Indo-European root *-twerkw*, which also yields the German *quer* (transverse), Latin *torquere* (to twist), English *athwart*" (Sedgwick 1993a, 12). To work athwart is, for Sedgwick, to be within the spheres of "continuing moment, movement, motive—recurrent, eddying, *troublant*." In Helmreich, it is to work "the empirical transversely," to recognize that "theories constantly cut across and complicate our descriptive paths as we navigate forward in the 'real' world" (2011, 134). As a sometime gender and queer theorist now all at sea with fish, I hold on to these different but compatible notions of athwartness. As will become apparent, I want to bring together, however awkwardly or athwartly, the worlds of fish and my background in feminist cultural studies.

Oceanic Feelings

The worlds of words and ideas continue to lure, and indeed the oceanic is replete with a paradoxical human desire to both make meaning of and be caught in the mystery of the oceans. Christopher Connery's (1996) article introduced to me the idea of "oceanic feelings," which has reverberated ever since. Connery reports how the phrase came into being. The writer and devotee of Eastern mysticism Romain Rolland apparently wrote to Freud on reading his *The Future of an Illusion*. We only have Freud's report on Rolland's thoughts, which he paraphrases: "It is a feeling which he would like to call a sensation of *eternity*, a feeling as of something limitless, unbounded, something 'oceanic.' . . . It is a feeling of an indissoluble bond, of being one with the external world as a whole" (Freud 2002, 2).

For Freud, it summoned early childhood, the unbounded that will later begin to be corralled in the constitution of subjectivity. I use it here to invoke the mystery and the majesty of the ocean. From within, floating with or looking toward it, the vision of the sea pulls. It offers an oblique entry into what might be an affective habitus of the more-than-human, of the oceanic as an inspiration for new forms of being. With this in mind, I turn to the question of the forms of representation and of their affective qualities we might mine for ways in which to render the ocean, its fish, and its workers as matters of care in public discourse. Here I gather together images of oceanic affects so as to unsettle common distinctions between land and sea, human and divine, fish and man, mind and body. This is in part a quest for the possibility of the oceanic and the oceanic as possibility.

The opening quotation taken from Coleridge's "The Rime of the Ancient Mariner" brings forth a very different disposition: "Alone, alone, all, all alone / Alone on a wide wide sea." As a child I loved reciting this poem with its menacing overtones.

> It is an ancient Mariner,
> And he stoppeth one of three.
> "By thy long grey beard and glittering eye,
> Now wherefore stopp'st thou me?" (Coleridge 1834)

Here it is man damned in his isolation from the sea, cursed for his act of violence and treachery toward the birds of the sea. The mariner kills an albatross—a well-known superstition among seafarers is that killing a seabird brings ill omen, because they are thought to be lost souls wandering the sea. Many read the poem as another example of the eighteenth-century romantic "awe of nature" and a gradual indication of "the consideration of animals as sentient beings" (Dwyer 2005, 14). However, the lines I chose to cite portray the stillness of solitude and human isolation in and on the ocean. The mariner is temporally and physically stranded in and on an indifferent sea. He is of course damned for his deed, and even after he returns to land he is condemned to tell his tale again and again with no resolution. He and the Wedding Guest are locked within the telling, the latter unable to physically escape just as the former cannot be freed from his burden. It is, if you like, a Sisyphean enactment where the end augurs the beginning.

I take these two dispositions, on the one hand expansive unboundedness as possibility and on the other the claustrophobia of isolation, as a springboard to other possible ways of incorporating the sea. Philip Steinberg's historical geography of the ocean gives depth to our current understandings of marine space. He takes inspiration from the thoughts of Strabo, the Greek geographer. Writing some two thousand years ago, Strabo saw in the ocean a space of comingling with human beings: "We are in a certain sense amphibious, not exclusively connected with the land but with the sea as well" (Steinberg 1999b, 368). Steinberg focuses on the flows of capital across time and space, and he identifies how the maritime became associated with fixity and with stasis. From the mid-eighteenth century with the focus on terrestrial industrial development, "the ocean became discursively constructed as removed from society and the terrestrial places of progress, civilization and development" (Steinberg 1999a, 409). Tracing the emergence of modern forms of regulation—quite surprisingly recent when one considers that UNCLOS (the United Nations Convention on the Law of the Sea) wasn't ratified in international law until 1994—Steinberg argues that the dominant understanding of the ocean is as "annihilated space," "a site of alterity," and as the domain over which capital is projected in its search for resources. Of course mercantile marine commerce and transport is still a constant, and as Kimberly Peters reminds us, "ninety-five percent of trade is still by ship" (2010, 1260). It is strange to think how slow the progress in hydrodynamics is. Steinberg writes that cargo ships still travel at the same speed as they did at the end of World War I (years ago, as I waited for my belongings to come by sea from Quebec to Australia, I did ponder that it took about the same time as did the ship on which my great-uncle worked taking wounded Australian soldiers home from England in 1915).

Steinberg is particularly damning of the nostalgic representation of the maritime as featured in countless harbor reconstructions around the world formulated for tourist tastes and not for the working boats and men, now relegated to being objects of tourism. He is also concerned that any fledgling discourse on the sustainability of the oceans is stymied by Hollywood "images of the ocean as devoid of nature, or as something to move through"

(Steinberg 1999a, 406). Steinberg wrote this before the phenomenal success of *Finding Nemo* (2003), one outcome of which was to momentarily stop kids from eating fish fingers. Contrary to Steinberg's pessimism, Peters argues that "maritime worlds open up new experimental dimensions and forms of representation." Of course, Paul Gilroy's brilliant book *The Black Atlantic* broke new ground in refiguring routes of human slave trade as well as the movement of black literary representation across the Atlantic, reminding us that "ships are living, micro-cultural, micro political systems" (1993, 15). To this we can add Sandro Mezzadra and Brett Neilson's argument, which, following Anna Tsing's work, looks to the "life and labor in these [marine] sites, where the boundaries between legal and illegal, licit and illicit, are often blurred and the nested scales of local, national, regional, and global no longer hold tight" (Mezzadra and Neilson 2013, 236). In Michel Foucault's words, "In civilizations without boats, dreams dry up" (1986, 27).

Pálsson uses the term "entangled resources" to focus on "the intimate relations of porous bodies and molecular environments" and to ask, "How should human-environmental relations be conceptualized and refashioned?" (Pálsson et al. 2013). Here the ocean provides us with a powerful horizon. I use horizon in the commonsense understanding as the line formed between sky and earth. In figure 1.3, it is intersected by what is apparently the longest jetty in Australia (or maybe the world, as it meets the ever-receding horizon to continue forever). The more specialized sense of horizon comes from the philosopher of hermeneutics Hans-Georg Gadamer. For Gadamer, the identification of horizon is integral to the processes of describing and interpreting: "The horizon is the range of vision that includes everything that can be seen from a particular vantage point. . . . A person who has no horizon is a man who does not see far enough and hence overvalues what is nearest to him. On the other hand, 'to have an horizon' means not being limited to what is nearby, but to being able to see beyond it" (Gadamer 1997, 302).

Gadamer's central point is that we need to be reflexive about what constitutes the horizon at any given moment. In this sense, much of contemporary cultural thought has, consciously or not, related horizon to the ground beneath our feet. Looking to the ocean as horizon promises to reorient our ideas. Of course, the ocean has long been a source of fascination for humans, but this has been in part because it seems so sublimely indifferent

Fig. 1.3. The longest jetty in South Australia at Port Germein runs into the horizon. Photograph by author.

to our wishes. As Gaston Bachelard recognized, "Water is truly the transitory element. It is the essential ontological metamorphosis between heaven and earth." But equally the sea "is inhuman water, in that it fails in the first duty of every revered element, which is to serve man directly" (quoted in Connery 1996, 291). I think this supreme disinterest of the ocean in human life is what Barthes is getting at when in a cryptic footnote in *Mythologies* he writes, "Here I am, before the sea; it is true that it bears no message" (1972, 112, fn2). Against the man-made significance of *la plage*, the ocean itself seems so unworldly, so foreign to us landlubbers that we cannot turn it into facile meaning. It cannot be scribed. Connery counters Barthes's denial of signification to the sea: "Yet signify it does, although in a manner beyond resolve. Is it the void that activates the terrestrial symbolic system? Is it the real beneath the floating discontinuousness of land; a symbolic system?" (Connery 1996, 290).

Seasick

If the ocean can provide and has provided a powerful horizon in reflecting on human alterity to our marine environment, there are associated affects that may provide new crucial methodological directions for more-than-human research. As I have touched on above, the ocean summons up diffuse affects, for which we do not always have the right words. From his fieldwork with Icelandic fishers, Pálsson's use of seasickness points to the physical and connotative upheavals that being on the sea can occasion: "Icelanders implicitly recognize the relationship between knowledge and practice, and the unity of emotion and cognition, body and mind. For them, 'seasickness' (*sjoveiki*) not only recalls the bodily state of nausea sometimes caused by the lack of practical knowledge, the unexpected rocking movements of the world, but it is also used as a metaphor for learning in the company of others" (1994, 901).

This moment of fundamental queasiness in the world—he himself physically experienced seasickness while conducting his fieldwork onboard—provides an embodied lens through which we can consider the different layers of ontological disordering of any posited distinct and separate entity: ocean-land, science-poetry, human-fish. As we all know, affect is embodied—to be alive is to be embodied in the world, and thus open to being affected and affecting others. But there are many forms of affect and many different ways to be embodied. While not an obvious affect, being seasick alerts us to the viscerality of being embodied and entangled in the swaying human and nonhuman nets of materiality and meaning. The sea, its movements; techne as learning within the pressures of the folds of human technology, the type of boat, the fishing gear; the tacit levels through which the past is reproduced in present practices—all caught in the prism of the rocking boat and the moving horizon.

This scene is also one where we learn with the elements that awaken us to our sheer lack of mastery before the might of the ocean. The form this learning takes is what Pálsson calls "enskilment," "a necessarily collective enterprise—involving whole persons, social relations, and communities of practice." For those who work on the sea as fishers, enskilment is learned in the presence of others, both human and most importantly nonhuman. This is embodied learning within the folds of the ocean brings together various forms of knowledge—past, present, cellular, felt, smelled, moved with, and so on. To give this ensemble of modes of learning its proper name, this is the realm of the tacit. This knowledge is unsaid; it is the undidactic, the learning from feeling. It is affect as embodied disposition. Perhaps it is not surprising that Pálsson also sees "enskilment in fieldwork, [which] inevitably involves psychosomatic processes, if not veritable 'gut reactions'" (Pálsson 1994, 902; see also Probyn 2004b).

To return to my central contention that we need to desimplify the sea and engage viscerally with her multiple elements, recognizing fishers' incorporation of the oceanic is important in several registers. For a start, as Kay Milton argues, "if people can identify with aspects of their ecological environment as being 'like' themselves . . . they are more likely to treat that environment as they might themselves or another person" (Nightingale 2013, 138). While this claim would be greeted in some quarters as yet more "human exceptionalism" (Plumwood 2007), the quest to formulate the more-than-human needs to acknowledge—as it pushes at—the limitations of human imagination.[6] In other words, if we want to generate care for the ocean and for her inhabitants, we need to work with the deep entanglement that fish, fishers, and ocean have forged over the millennia. Andrea Nightingale's work on the subjectivity of fishers furthers this line of inquiry into the embodied subjectivity, or habitus, of fishers. Her research looks particularly at the embodied divide between those who fish and those who regulate fishing. At heart this is about whose knowledge counts. Nightingale focuses on how and where subjectivities emerge. As she succinctly puts it, "Fishing produces particular kinds of subjects and bodies" (Nightingale 2013, 138). She elaborates: "It is the embodied act of working on wet, smelly, cold and dangerous boats that is important in creating a boundary between the subject of the 'fisherman,' 'community,' and the 'sea.' . . . The sea is the defining feature of their lives. . . . A sense of self is shaped by the sea" (142). Not only are fishers shaped by the sea, they are increasingly squeezed between different frames, which position them as "the exploiter of the sea" (143). Or in the framing of the documentary *Sea the Truth*, they are the "destroyer of ecosystems" (Everaert and Zwanikken 2010).

My questions remain: What forms of care are most effective in changing our behavior? How do we come to care? How can we make questions of fish and humans visceral? Trying to connect between the more cultural and the more empirical dimensions, I have attempted to model a more-than-human entanglement that is expansive and affecting. It is a fraught assemblage. Michael K. Goodman reminds us of "how absolutely viscerally entangled food is in the landscapes of contemporary capitalistic political economies" (2008, 5). Here he foregrounds visceral affect with and within the economic. This problematizes Branson's claim that "the consumer can make a difference." Of course we can, and the many campaigns are having a partial, uneven, and gradual impact on what people choose in the supermarket. We also need to qualify which consumers. As a middle-class woman living close to decent supermarkets and within walking distance of the third-largest fish market in the world, I can make choices because there are things to choose from, and I have the money to be able to buy what I deem beneficial to the fishers, fish stocks, oceans, and my body. I spend a lot of time peering at the labels on tinned fish and asking fishmongers questions. While in the best of all possible worlds we would all have the resources to do this, we simply don't. The onus on consumer choice is so often a neoliberal fantasy, and the fork as "a mighty weapon" (as Thieme claims in *Sea the Truth*) is a very one-dimensional tool (Everaert and Zwanikken 2010). It is often a weapon wielded against others—those who are seen as the enemy, those who simply do not have the economic or cultural capital to be able to choose.

In these scenarios, more-than-human caring becomes a zero-sum game. Ethics becomes moralistic black and white. Instead of such thinking, we need to conjugate the multiple relations between and among ocean, fish, and people. As Mezzadra and Neilson write, this implies "a process that simultaneously folds and unfolds spaces . . . [that reveals] new regional, continental, and transcontinental routes of connection." The multiple matters of relatedness of fish and humans further contribute to what they call the "uncanny stretching and overlapping of geographies" (2013, 212).

While for some an individual choice may be to just say no to fish, there is evidence that doing so lessens people's attachment to commonly held goods. This is definitely not to replay Garrett Hardin's (1968) "tragedy of the commons," to imply that it is only under socialism or free enterprise that the commons can be saved. It is less about self-interest and rather more about more complicated attachments we form to those that procure and prepare what we eat. Those entanglements are simultaneously economic, as well as geographic, and they are also allowed for and colored by our affective habitus. To return to Nightingale's research on Scottish inshore fishermen, she describes how they "see themselves as 'part' of the sea." They are "attached to the sea . . . and attached to a sense of treating the sea well" (2013, 142, 144). They are strongly opposed to the large trawlers, seemingly

Fig. 1.4. Statue of boy with bird at the port of Peterhead, Scotland. Photograph by author.

not because of competition but because "it's a business as opposed to a way of living" (145). These fine distinctions are crucial, and they are afforded by an oceanic habitus.

The Temporality of Caring

A young fisher lad throws leftover bait to a gull. They are caught forever in this gesture while another gull flies by squawking, more interested in live bait than in statues (figure 1.4). The boy looks over the fish market at Peterhead, which was a flourishing place until the battle among fishing nations ended its prosperity. The fish that they fought over are long gone, but the fisher lad remains, a reminder of what was.

This chapter has sought to bring together several differing representations of the relations between humans, fish, and ocean. The image I used at the outset was that of a clapotis wave. This happens when waves break onto structures and form into what are called standing waves. As kayakers know, this can be scary. It's easy to lose one's direction. In this chapter there are deep and maybe intractable divides that I have paddled over too quickly in my quest to frame an oceanic cultural studies that is more inclusive of many and sometimes warring concerns. I am sympathetic to Mentz's conceptual gamble on the "post sustainable," as I am to Alaimo's argument that "the tendency of sustainability [is] to externalize and objectify the world though management systems and technological fixes" (2012b, 563). However, in this book I hope to move beyond reproducing impasses. In particular I want to figure a different kind of relating of people, oceans, and fish. This is to translate across various concerns, different ontologies and epistemologies, and different ways of enacting a sustainable ethics of more-than-human fish. How and whether and even if sustainability needs to be sutured to the future (as well as how that is figured and imagined) remain important questions. In the case studies of this book, we will find evidence of how sustainability as futurity works or doesn't work as a set of practices. This depends greatly on the timing of sustainability, of how we conjure its tenses as a verb—to sustain. In turn, this requires that we rethink time and timing.

Posing sustainability as a fixed horizon places it as a form of codified or codifiable moralism. This is an anathema for the ethical thinking that Gilles Deleuze recovers from Spinoza. As I alluded to in the introduction, Deleuze elaborates on animal ethologies such as those of Jakob von Uexküll to formulate a Spinozan ethics that defines bodies "by the affects they are capable of" (Deleuze 1992, 627). As Deleuze writes, "That is why Spinoza calls out to us in the way he does: you do not know beforehand what good or bad you are capable of; you do not know beforehand what a mind or body can do, in a given encounter, a given arrangement, a given combination" (1992, 627). Simplifying the sea in cultural representations and arguments assumes we already know who the culprit is and what the outcomes of prescribed actions are. This is not ethics; it is a moralism that confines human thought and practice.

Moira Gatens extends Deleuze's ideas into a feminist politics or ethics where action or decisions are predicated on timing. She writes, "Ethics is here not conceived as a transcendentally guaranteed set of rules but an ongoing experiment that requires skill, patience, and great care if it is to turn out well" (Gatens 1996, 11). That skill, patience, and care are what is lacking in contemporary debates. It is fussy, never-ending work that is most often to be found in the realm of the domestic and the feminine. In Gatens's thinking, it is explicitly feminist. It involves a different sense of temporality, one that insists on acute judgment of timing. It cannot be focused on an abstract futurity. Extending Deleuze's argument, Gatens contrasts two Greek words for time and timing. *Chronos* is about measuring time in ways that are deterministic. *Kairos*, however, is about being in tune with the flux of assemblages, waiting for the moment that presents itself as "only at 'this' time" (Gatens 1996, 15–16).

This sensibility resonates with what Sarah Ensor calls a "spinster ecology" (2012, 409). Ensor relates how a former secretary of agriculture under Dwight D. Eisenhower sought to discredit Rachel Carson. He asked "why a spinster with no children was so concerned about genetics" (Ensor 2012, 409). Ensor points out how misogynist the statement was but also, more tellingly, how widely and powerfully the trope of children is deployed in discussions of sustainability: a "rhetoric predicated on matters of inheritance and procreation" (409). This rhetoric can be found across any number of disparate sites. As I argue one of the more ironic examples can be seen in WWF's slogan "The future is man-made."

Ensor's insight about the problematic futurity of sustainability discourses is about timing. It is not the temporality that is so often conveyed through discourses of sustainability—save the world, save the fish, for your grandkids. Of course many people, including queers like myself, will not have grandchildren to whom we are supposedly bequeathing our efforts to save the oceans. More importantly, this injunction frames the complex issues of sustainability or of fish and human relations as a heteronormative end goal. It simplifies, strips away detail, complexity, and the beauty of intricate human-fish relations. I question whether this type of exhortation is the best way of getting anyone (parents, grandparents, aunts, or not) to care about what is happening within the realms of ocean-fish-human entanglement. This framing of the time of sustainability yokes a stable present to an unknown but presumably remediable future. This reinforces the individualist stance of the "I vote with my fork" mentality that I critique, and endows the human subject with choice and complete agency. This is so at odds with the complex and interwoven sets of practices of sea, people, and fish.

As I have sketched out here, the ability to care in nuanced ways comes in different forms. It may be through a habitus that is open to the ocean, affected by the millennia of human awe and engagement. It may be through feeling a connection, however extenuated, with those who have the sea in their blood. It may be through recognizing how our bodies taste of salt. It will not be, however, through moralistic directives. It will not be through simplifying the sea. It may be that through coming to care we become attached with others, and imbricated with the movements of the ocean. A cultural politics of more-than-human marine entanglement must, I think, be open to the rocking relationships of people, fish, and ocean. It is to these attachments that I turn in chapter 2, where we encounter the storied oyster in all its exquisite tastes, smells, textures, and relations.

2

Following Oysters, Relating Taste

Oyster Encounters

I've been following oysters for some time. Several years ago, my Scottish girlfriend and I sat in Rogano in Glasgow, a wonderful art deco fish restaurant that opened in 1935—the same year that the *Queen Mary* was being built on the Clyde. In the pale Scottish sun, we devoured oysters so fresh from the sea that we forgot the sounds and smells of inner-city Glasgow. "Where do they come from?" we asked the young waitress. "Och, that's easy. They're from Loch Fyne," as she pointed vaguely north and west. The taste memory of that perfect oyster sent us in search of a small loch in western Scotland, a tale to which I will return.

In her book singularly titled *Oyster*, Rebecca Stott remarks, "The history of the oyster-human encounter is a history characterized by intimacy and

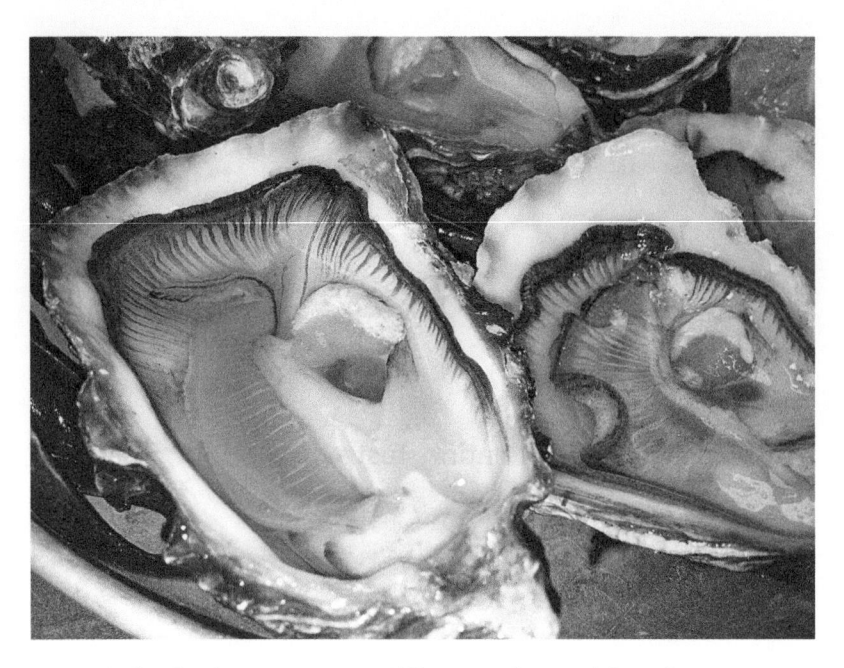

Fig. 2.1. Perles Blanches no. 3 oysters at L'Ilôt, Paris. Photograph by author.

distance" (2004, 10). I like the notion of an encounter, an encountering of oyster and human—the pull and push of intimacy and distance, of desire and disgust. As an oyster eater I am always astonished at the force with which those who don't like oysters express their distaste. I am totally bemused by those who pretend to be indifferent to oysters.

In this chapter, I follow oysters, I encounter them, I eat and relate what it might mean to become an oyster-eating body in training. To parse these verbs, we will follow oysters into their history and milieu. This is both akin to and different from Ian Cook et al.'s (2004) now-well-known method within cultural geographies of food. There the objective is to "follow the thing" back into its intricate imbrication of agricultural origins, producers, and supply chains. Like Cook, I advocate an active practice of following, but here it is less about following them into supply chains (although, as we will see, the global-local market connections of people and oysters is a rich site); rather, I follow them into realms of taste and history. We encounter them in their deeply cultured and storied milieu. This is a more rhizomatic following, where the encounters send off shoots in all directions. And finally, I eat an oyster and relate this to what oyster eating does.

Tonguing and tasting an oyster is something you do mindfully with all your body. It is a "perplexed" activity (Hennion 2007), which is to say, it makes you on the lookout for what it is doing to you, or what it is doing to others. You may be wary of the bad oyster. You may be thinking that this one is zinckier, or creamier, or brinier than the previous one and maybe the one to come. You may have been memorably told, "All I want are oysters, your lips, and Sancerre." You may wonder whether to bite and chew—how many times? You may experience that in the tasting of the oyster your body reverberates all over. The slide of an oyster down your throat may inaugurate a supremely reflexive process where one observes oneself tasting: I am eating this Sydney rock oyster from Merimbula on this rainy late winter Sunday in Sydney. There are at least three of us present: her, the oyster, and me. What sorts of relations are forged in this instance? The "we" eating is of course problematic on several fronts: As Simmel puts it, "What the individual eats no one else can eat under any circumstance" (1994, 346). For Simmel, this paradox lies at the heart of human sociability—perhaps a more-than-human sociability. We have to eat, and in so doing we hope to bridge the chasm that lies between us.

In chapter 1, I argued that oceanic relations are necessarily complex. I want to disturb how the sea is simplified, perhaps especially in certain forms of ocean and fish activism. If we are going to eat the ocean, with all the devastation and pleasure that the phrase holds, we must do it carefully, to cite Mara Miele and Adrian Evans (2010). It is important that each time we eat the ocean, our senses and our sense of relatedness are heightened. As we saw in chapter 1, marine scientists are concerned about how human activities are producing severe anthropocenic changes in oceanic ecosystems. They warn that this is producing a stripped-down and simplified sea. My argument is against a social and cultural simplification—the social and public discourses that are producing a simplified understanding of what we, as consumers, can do. Thus I critique the notion that our forks are mighty weapons that we can wield for a sustainable future. The "our" invoked in these discourses is misleadingly inclusive and tries to galvanize the image of a unified mass of consumers, forks in the air, striding into a sustainable future. This ignores the millions who depend on fish. Ignorance of a wider, more complex ocean radically simplifies an ethics of connection with the sea, her inhabitants, and her dependents.

Thomas Henry Huxley called the oyster "a delicious flash of gustatory lightning" (Hardy 2003, 72). I am drawn precisely to this most emblematic of mollusks because of how it relates to so many realms, scrambling border distinctions: cultured and storied, nakedly rubbing flesh on flesh, nature to nature as it caresses the insides of my throat. But I am equally drawn to its own resolute nature. Witness and participant throughout human history and beyond, extant since 200 million BC—it remains obtuse.[1] Its shell is the first and sometimes the only thing that humans and other predators encounter. Hard enclosed, the oyster does not give itself willingly or easily.

I eat an oyster . . . the oyster eats me. In her argument, Annemarie Mol writes that "to say 'I eat an apple' evokes a particular situation" (2008, 28). She uses this phrase and her act of eating an apple to otherly conjure relations of subjectivity. Marilyn Strathern closely reads Mol's argument and writes, "Subjectivity, in this rendering, is initially posed as at once a question about a state of being and a question about the agent who takes action. To focus on 'eating' dislocates the two, insofar as the eating self is not an agent in any obvious sense, as Mol says" (Strathern 2012, 3). This obviously complicates any simplified politics of eating. She continues, "An eater cannot help being a subject, although who or what may be problematic" (5). I take the oyster as a case in Mol's sense: "A case is something to explore, to learn from. Attending to it carefully may make you reconsider what you thought was clear and distinct. It may interfere with your very language. And while a case cannot be generalized, neither is it local. Instead, its specificities are made to travel" (2008, 32). Elsewhere Mol and her team use the generative phrase "matter-related" (Abrahamsson et al. 2015, 9). They argue that in the face of a barrage of arguments about "matter," "materiality," and "the vitality of matter," we miss the fact that matter is never matter by itself. As they say, fish "is not 'fish itself,' but 'fish-related' too. It relates to the environmental and climatic changes that transform its habitat. It relates to the zooplankton and smaller fish it eats, and to the other species (including human beings from various parts of the world) that feed on it" (Abrahamsson et al. 2015, 9).

I encounter an oyster; I eat an oyster: The oyster always tells from whence it comes. Oysters are as close as most of us get to eating the sea. They taste of "merroir"—as with the term "terroir," this designates the physical factors of place, techne, and tide that give each microenvironment a different flavor. As Rowan Jacobsen writes, "The comparison of oysters is . . . not another food but always a place . . . the sea" (2008, 2). Oysters give us a taste of "somewhereness" (Jacobsen 2008, 3). In c. 138 BC, the Roman satirist Gaius Lucilius wrote, "When I but see the oyster's shell, I look and recognize the river, marsh or mud, where it was raised" (Yonge 1960, 17).

As well as always being emplaced, situated within their ecosystems, oysters frame a powerful temporality. They tell of history, and as entities they have been caught within ebbs and flows. The Romans loved oysters and were the first to farm them. Oysters have been caught within the all-too-regular boom and bust occasioned by rapacious human appetites. They have been dredged to extinction, and yet in different forms they return. As he sheltered from Hurricane Sandy in 2012, Paul Greenberg (2012) reflected on the pivotal role that oysters had once played in protecting New Yorkers from storm surges, with a "population that numbered in the trillions and that played a critical role in stabilizing the shoreline from Washington to Boston."

This is part of the doing of oysters. An oyster is a powerful keystone species: Each adult oyster filters and cleans up to fifty gallons of water per day—gobbling up phytoplankton and removing dirt and nitrogen pollution. As we saw in chapter 1, excess nitrogen in seawater can build to the extent that the water becomes drained of oxygen. One of the great oyster grounds in North America is Chesapeake Bay ("great shellfish bay" in Algonquin) on the Eastern Seaboard. The population has been decimated (down 98 percent), but in precontact times the vast numbers of oysters could clean the entire eighteen trillion gallons of water in the bay in a matter of days (Chesapeake Bay Foundation 2014). The website of the Chesapeake Bay Foundation, set up to re-create other oyster environments in the hopes of cleansing the bay, features a time-lapse short video where the oysters lie in near invisibility due to the pollution. In minutes the scene is pristine, revealing the interaction of oyster, reef, algae, and plants as well as so many other entities impossible to see with human eyes. Oysters are so important to humans, but, equally if not more, they are deeply related to their milieu. It is this particular set of qualities I want to relate here.

And then, of course, there's oysters and sex. Oysters are very queer. Their reputation as an aphrodisiac has been long vaunted, and we know the tales of Casanova's prodigious appetites for both oysters and sex—fifty for breakfast got him started. In the West, humans project female genitalia onto oysters, morphing in culture the bivalve and the vulva. This may explain why they occasion such vehement reactions for or against. The scientific explanation for oysters' long reputation as an aphrodisiac is that the mollusks are rich in zinc, iron, calcium, and selenium, with healthy levels of vitamins A and B as well. Oysters also have the amino acids that trigger increased levels of both testosterone and progesterone ("Time to Shell Out!" 2014).

Personally I think it's a bit more complicated than that; it's how they complicate our relationship of human to nonhuman to human that is so queer. Sarah Waters's (1998) exciting romp through Victorian England, *Tipping the Velvet*, makes this very clear. "Tipping the velvet" is a Victorian euphemism for cunnilingus (Kurth 1999). In the words of Nan, Waters's oyster-loving lesbian protagonist, "Oyster-juice [is] my medium. . . . I never doubted my own oysterish sympathies" (1998, 4).

The oyster is the emblem of the in-between: growing happily in brackish water lapped by sweet and saline waters, changing its sex frequently. But the analogies between oyster sex and human gender are ham-fisted. For instance, in Trevor Kincaid's simplification, "With the oysters it's strictly a case of Dr. Jekyll and Mrs. Hyde. . . . Last spring these little fellows were sitting around talking baseball and politics. Now they're all knitting on tiny garments" (Steele 1964, viii). Against Kincaid's awkward relating of humans and oysters (he was after all a famed professor of zoology, not literature), the great American food writer M. F. K. (Mary Francis Kennedy) Fisher rendered the oyster in all its queer glory (she was so much more than what "food writer" usually signifies[2]). In her chapter "Love and Death among the Molluscs," she writes, "Almost any normal oyster never knows from one year to the next whether he is he or she, and may start at any moment, after the first year, to lay eggs where before he spent his sexual energies in being exceptionally masculine. If he is a she, her energies are equally feminine, so that in a single summer, if all goes well, and the temperature of the water is somewhere around or above seventy degrees, she may spawn several hundred million eggs, fifteen to one hundred million at a time, with commendable pride" (Fisher 1990, 125).

It is not all fun and games, however. Oysters have many specialized enemies. Apart from rapacious human appetites, there are "Drills or Tingles or Sting Winkles [that] bore circular holes in the shells of oysters with their ribbon of tiny horny teeth." They then secrete a lime-dissolving liquid and "literally eat the oyster alive, scooping out its flesh with their proboscis" (Yonge 1960, xiii). Killer starfish are even more horrific in their quest to devour the oyster. With their feet they pull the oyster bivalves apart (quite a feat, as anyone who has tried to open an oyster with her bare hands will attest), and then they evert their stomachs (turn them inside out) and squish them through a tiny crack and digest the oyster's living tissues by absorbing the ooze directly into their bloodstream. "And then they retract their stomachs" (Yonge 1960, xiv). M. F. K. Fisher describes the scene thus: "The starfish . . . floats hungrily in all the Eastern tides and at last wraps arms around the oyster like a hideous lover and forces its shells apart steadily and then thrusts his stomach into it and digests it. . . . The picture is ugly" (1990, 127).

The conceptual tension here is how to pry open the oyster without killing it—to frame it in its own materiality while at the same time reflecting on its commensalist role. In ecology theory, commensalism refers to a relationship in which one organism benefits from the other (Hogan 2012). The proviso is that the organism isn't harmed. And of course eating cannot be said not to harm the oyster. But might there still be ways to consider the coconstitutional relatedness of oysters and humans? As M. F. K. Fisher astutely puts it, "The oyster has eight enemies, not counting man who is the greatest, since he protects her from the others only to eat her himself" (1990, 127). Fisher continues about human-oyster relations, which are close if far from altruistic: "Men have enjoyed eating oysters since they were not much more than monkeys . . . and thus, in their own one-minded way, they have spent time and thought and money on the problems of how to protect oysters from the suckers and borers and starvers" (128).

Enclosed and recalcitrant, the oyster directs a new way of thinking that mines its culturedness while ruffling the all-too-human propensity to remake it in our desires. In the realm of the athwart, the oyster queers the divide between the human and the nonhuman, eroding human exceptionalism while recognizing the traces of our corelatedness.

Oysters, wild or not, are cultured entities. In Australia, oyster middens have been dated showing that ten thousand years ago oysters were an important part of coastal Aboriginal people's everyday life. Much like middens, the landscapes of our minds and histories and those beneath our feet are littered with examples of this mixed materiality. In a quest to rethink "thing" connections without sacrificing the necessary otherworldliness of human-thing fabrications, Michael Parrish Lee turns to the weird and novel conjugation of things to food to animals in Lewis Carroll's *Alice's Adventures in Wonderland* (1865). Lee argues that in Carroll's writing, "things are not objects that appear uncannily human, but edible life-forms that presumably have appetites of their own" (2014, 486). In "The Walrus and the Carpenter," we encounter the oyster at the epicenter of appetites. Although this is often cited, let us take pleasure once again in Carroll's relating of oysters.

"O Oysters, come and walk with us!"
The Walrus did beseech.
"A pleasant walk, a pleasant talk,
Along the briny beach:
We cannot do with more than four,
To give a hand to each."

. . .

But four young Oysters hurried up,
All eager for the treat:
Their coats were brushed, their faces washed,
Their shoes were clean and neat—
And this was odd, because,
you know,
They hadn't any feet.

. . .

The Walrus and the Carpenter
Walked on a mile or so,
And then they rested on a rock
Conveniently low:
And all the little Oysters stood
And waited in a row.

. . .

"A loaf of bread," the Walrus said,
"Is what we chiefly need:
Pepper and vinegar besides
Are very good indeed—
Now if you're ready, Oysters dear,
We can begin to feed.

. . .

"It seems a shame," the Walrus said,
"To play them such a trick,
After we've brought them out so far,
And made them trot so quick!"
The Carpenter said nothing but
"The butter's spread too thick!"

"I weep for you," the Walrus said:
"I deeply sympathize."
. . .

"O Oysters," said the Carpenter,
"You've had a pleasant run!
Shall we be trotting home again?"
But answer came there none—
And this was scarcely odd, because
They'd eaten every one." (Carroll 1872)

The tale of the poor oysters in "The Walrus and the Carpenter" had a profound effect on the English public. A cartoon published in *Punch* in 1903 depicts the Walrus and the Carpenter clutching their stomachs, as around them are strewn empty oyster shells, with the title "AVENGED!" The text reads:

"O Carpenter," the Walrus said,
    "I sympathise with you.
You say that you feel rather odd,
    I doubt not that you do.
for, curious as it may appear,
    I feel peculiar too.". . .

"O oysters!" moaned the Carpenter,
    And that was all he said,
As on the coolest piece of rock
    He laid his aching head.
The Walrus, too, refrained from speech,
    He was already dead.[3]

Avenged indeed. At the turn of the last century there was great concern about the link between oysters and shellfish and typhoid. The *Punch* cartoon appeared shortly after four people died at a banquet in Winchester, where the new discipline of epidemiology had ascertained an oyster and enteric fever etiology (Morabia and Hardy 2005). They were troubled times for oysters. Poor hygiene and pollution rendered them ideal carriers of disease. Imagine what the Thames was like in the late nineteenth century. Following what was called "the great stink of London" in 1858, plans began for the modern sewage system (Halliday 1999). Nonetheless the Thames was basically the main sewer of London, and it would have been vile. As J. Donald Hughes writes, "Sewage might flow down the river during low tide, but twice a day a wall of water would carry it back upstream" (2001, 120). We know that oysters try their best to filter and clean, but washed in sewage several times a day, their constitution would be no match for disease. In addition, the actual oyster beds were fast vanishing because of industrialization. As Rebecca Stott recounts, Lewis Carroll may have had in mind the dire state of English oysters when he wrote his account of the greed that gobbled the young oysters. In 1871 *The Times* decried the overfishing and consumption that had caused the decimation of native oyster beds and the concomitant hike that forced the price of oysters well beyond reach of the poor, who had looked to them as subsistence food. The piece in *The Times* is so very prescient of a now-common state of affairs. "From prehistoric man to August, 1864, is a long stretch of time . . . and what has been done in this period? Why, all the Oysters in the sea, or at least in our seas, have been eaten up. The Oysters are all gone and no wonder. . . . The Oysters cannot run away from their destroyers, nor be induced to come back again when their persecution is over" (Stott 2004, 93).

Taste Relations

I feel a ghostly, sympathetic twinge in my wrist and finger-joints at the sight of a fishmonger's barrel, or the sound of an oyster-man's cry; and still, sometimes, I believe I can catch the scent of liquor and brine beneath my thumb-nail, and in the creases of my palm.—Sarah Waters, *Tipping the Velvet*

Oysters compel a radical rethinking of our notions of tastes, taste, and tasting. In the above quotation, Sarah Waters has her heroine in *Tipping the Velvet* convey how deeply the taste for oysters pervades the body. You could say that her body has formed a cellular memory of oysters. I want to take this connection to experiment with ways of reconceptualizing taste, pushing it past its normal position as social marker. Taste is, of course, often considered a fundamental sign of one's class position in society, a mechanism that in Pierre Bourdieu's (1984) terms serves to reproduce inequalities. Taste, for Bourdieu, is the embodied mode through which hierarchies are affirmed: "Taste classifies and it classifies the classifier" (1984, 7), as his most famous of dictates has it. Bourdieu's points about taste as the exercise of judgment were based on his copious questionnaire data gathered in 1960s France, which queried people about their eating activities as well as other leisure practices. In the decades since Bourdieu's *Distinction* appeared, taste has tended to become reified as social casting. This leads to rather stultified statements such as "Eating is a socially constructed practice" (Wills et al. 2011, 726). There is no doubt that one's background, including familial social class and education, influences what one thinks tastes good. "For Bourdieu, individuals can no more 'step outside' the boundaries of their classed habitus than an 'outsider' can choose to step in to a completely different world in terms of taking up its associated 'alien' practices and habits" (Wills et al. 2011, 727). The frame of class remains crucial, but it is important to remember that the ability to appreciate oysters is not the province of the few. Oysters were protein for the poor. In Sydney in the mid-1800s, oysters were cheaper than eggs and milk, in part because anyone could collect them off the banks of Sydney Harbour.[4] In Sydney Cove there would have been dozens of eateries selling oysters, sometimes to former convicts who had been sent to the end of the world for stealing a loaf of bread in England.

How do you eat an oyster? Tonguing is what I do when with an index finger I propel the oyster into my mouth. I eat an oyster, and there is a moment of confusion or anticipation when I don't quite know what I am tasting: sea, flesh, memories. In the words of the French sociologist Antoine Hennion, "Taste is a problematic modality of attachment to the world" (2007, 101). Hennion reworks taste so that instead of being a static marker of difference, it is a mode of relating: "Taste is not an attribute, it is not a property (of a thing or of a person), it is an activity. You have to do something in order to listen to music, to drink a wine, to appreciate an object. Tastes are not given or determined, and their objects are not either" (101).

These descriptions begin to capture the way taste acts as a connector between history, place, things, and people. What I particularly like about this approach is that it foregrounds the mattering of taste. A British food blogger gets the idea. "Eating raw oysters is a uniquely invigorating experience; a bit like battery-licking for grown-ups. It seems that we can taste the elements they contain: zinc, calcium, copper, iodine, magnesium. And no other food conjures up a physical feature of the Earth as strongly as a bracing, salty, tangy oyster: the essence of the sea in edible form" (Eat the Seasons 2008).

This description conveys how the oyster-eating body is a body in training—stretching senses, flexing hidden muscles, firing synapses. It takes time to begin to identify the pulsations of eating oysters, to name the minerals, and the textures of the sea. The body encountering the oyster tries to sift through different sets of senses and to describe how they fit within a complex web of things. Bruno Latour writes that learning to be affected and "training the nose" means "acquiring a body [in] a progressive enterprise that produces at once a sensory medium *and* a sensitive world" (2004, 207). As the body becomes a sensitized medium, relations between different parts of the body vibrate differently in tandem with different aspects of the thing being tasted. This activity remakes the world, makes up new ways of relating. Latour talks about how odor training changes the relationship between words and their objects. To talk about tasting oysters as battery licking, as that blogger does, is to think differently about how body parts act and react to and with other nonhuman bodies. As Alison Hayes-Conroy and Jessica Hayes-Conroy frame it, this is "the visceral realm—the realm in which the whole molecular ensemble of the minded body feels the world, the realm from which life processes and events precipitate and hence in which political activation *materially unfolds* (or fails to)" (Hayes-Conroy and Hayes-Conroy 2008, 462).

M. F. K. Fisher's writing about eating exemplifies the visceral ways in which taste relates class and gender. This is particularly the case in her oyster writing, which brings together a collusion of sex, class, and bivalve, while at the same time remaining attentive to the materialities of the oyster. Listen, for instance, to her short story where class, taste, gender, sexuality, and even generational difference are made to relate through the taste of oysters.

"The First Oyster," Fisher's autobiographical tale of adolescent female sexuality and pashes in a California boarding school, was written when she herself was but a girl. As she builds her story, wending around and around the figure of Mrs. Cheever, the cook, who produces unappreciated marvels of well-prepared food for the young ladies, it is above all a love story in which a taste for oysters both marks and transcends class. Mary Francis tastes her first oyster: "Oysters, my delicate taste buds were telling me, oysters are *simply marvellous!* More! More!" (Fisher 1990, 375). One can picture the scene: The girls in a bright room, seated at several long tables. There is of course a bad girl. There always is. In M. F. K.'s story, it is Inez, an upper-class girl slightly older than Mary Francis, who has been making a play for the young M. F. K. The oyster still swelling in her, Mary Francis escapes her clutches and rushes into the kitchen. "The delightful taste of oyster in my mouth, my new-born gourmandise, sent me toward an unknown rather than a known sensuality" (376). There in the kitchen, she stumbles upon a tableau of tenderness between the cook, Mrs. Cheever, and the unnamed nurse. The cook with tears "running bloodlessly down her soft ravaged cheeks" serves the nurse—her *amie*—a platter of oysters, watching as "the old woman ate steadily, voluptuously, of the fat cold molluscs" (377). As I said, it is a tale of class order overturned by the taste for oysters. The teachers and students looked down on Mrs. Cheever—just a cook—and Mrs. Cheever affected disdain for the nurse as even lower on the pecking order. But through and with the oysters, M. F. K. affords a wrenching depiction of female care that grasps the two old women—perhaps lovers, perhaps long since just companions—in a kitchen eating oysters. I see them as Fisher does, focused on their love of eating oysters—not the remains shunned by privileged young girls.[5]

Working-class women vanquished by oysters, upper-class girls trying out their newly acquired tastes for other girls and for oysters. How many ways does the oyster initiate different forms of embodied entanglement and attachment? From its wondrous ability to shift sex and its intricate anatomy to the middens of history that tell of human-oyster interaction, the oyster encourages us to think large. In its privileged position, situated near the shore in the mixed waters where rivers flow with nutrients mingled in saline from the ocean and wash in and out of the oyster's cilia, it asks us to follow its relations to place and time, to the specificities of its whirls that spin together so many elements.

Following Taste

It is time to follow the oyster, to encounter it within its habitat, and thence to relate the shoots sent off in relations of intimacy and distance. To aid in this simple yet somewhat twisted movement, I look to a model of "*involved* description: ethnographic work that looks for contrasts, sets up differences and seeks for what one practice might learn from another" (Harbers, Mol, and Stollmeijer 2002, 219). Involved description is always situated description. It all starts (and ends) with place. But place is also written with history, and culture is written into the land, as Raymond Williams (1989) beautifully put it. It is also written into the sea. You can't understand the flourishing oyster industry in the Eyre Peninsula, and its isolated and rugged coastline in South Australia that faces Antarctica, without understanding the connections forged by history, migration, and the clusters of settlement that form and reform around water, shoals, harbors, and land. The first human occupants of the Eyre are the Nanuo Aboriginal mob, who would move between the sea and the land depending on the time of year.

The early white settlers were "peasant fishermen" from Ireland, fishermen and farmers. According to Paul, a retired fisherman I met in a pub in Coffin Bay, there used to be an ethos of "farming the sea rather than strip mining it." Over a couple of days, Paul told me about the area. Paul owned a rock lobster boat, a very lucrative business; nonetheless, he was deeply affected and invested in the history of oysters in the area. Over a couple of glasses of his favorite "chardy" at the local pub, he related the history of the disastrous effects of dredging, which killed off the native oysters. The angasi oysters (*Ostrea angasi*) were endemic to southern Australia. They liked sandy shallow waters and were easy prey for the newly arrived peasant fishermen. He recounted the complex interactions between different ethnicities who extracted food from the sea. In the mix were the Chinese gold miners who arrived in South Australia in the mid-1800s and made their way slowly by foot to the Victorian gold fields. They worked the land and the water and left traces of their culture on the Anglo-Celt population. In the pub, Paul pointed to a young fisherman who had a tattoo of the lucky 8 on his arm—the coiled snake signifying luck within Chinese culture. Just as the traces of those Chinese can still be seen, so too are the angasi—now being farmed along with the exotic Pacific gigas.

"Tae think again." This is the punch line in the unofficial Scottish anthem, "The Flower of Scotland," which tells of the Scottish victory over the English in 1334. Improbable as it seems, I want to use the spirit of this phrase ("Tae think again") and return to that perfect oyster (one of many perfect oysters, if such a thing is possible) we ate in Glasgow so many years ago now. Come with me as I follow it through to its beginnings in the west of Scotland. Loch Fyne Oysters lies at the top of the loch in the little village of Cairndow. The road we take twists and turns, and we gather our stomachs at a stopping point called Rest and Be Thankful. The hand of the Highland Clearances is everywhere—from the limited infrastructure and sparse population to the beautiful wilderness that is a testament to the lack of people, modernity, industry, and money. The Clearances are widely considered the worst disaster to hit the Highlands, in part because it was inflicted by the chiefs of clans who had been held in high regard for centuries. Following the failed Jacobite rising in 1745, the clans began to disintegrate, as did the system of obligations and responsibilities that had long held together Gaelic Scotland. The vast estates were run by absentee landlords—by their factors—while their masters lorded it down south in London. The factors were agents of the lairds, often brought in from the Lowlands to do the dirty work of throwing the peasants off the land. (My father used to say that my forebears were Lowland factors, but then he didn't like my mother's Scottish family.) Sheep were introduced because of the high price of wool. The landlords brought in workers from the south and rendered the small tenants redundant. For a time the locals eked out a living by harvesting kelp, but as the rule of the landlords and their factors became ever more draconian, they were forced to emigrate in poor circumstances to faraway places: to the "pink bits" that were the British Commonwealth.[6] "Whether it was economic necessity as described by some, or ethnic cleansing, as described by others, the net result was that between 1783 and 1881 man's inhumanity to man resulted in a documented 170,571 Highlanders being ejected from their traditional lands" (Scottish Tartans Authority 2014).

As the tone of this description attests, it is not an overstatement to say that Highlanders still live the legacy of that time. The Highland landscape is a stark example of the often-vexed interrelations of people, animals, and environment: It's a habitus, if you like, but one that includes the elements, weather and climate, a history of empty stomachs, fear, and respect for the sea and for nature's reproductions. There is always a thin but visceral layer of remembrance: that some humans in their greed placed the economic returns from wool higher than their fellow clan members. And of course some of those same people who were cleared from their ancestral homes came to Australia and may have participated in the reenactment of another hierarchy, this time over the Aboriginal inhabitants, placing them on par with the flora and fauna of this bewildering new land.

Lack of employment and the sea is key to the story of Loch Fyne Oysters, as was an enduring sense of the relations to community, land, and sea. This is where Johnny Noble enters the picture. Noble was the son of an industrialist, Sir Andrew, who had made the family fortune in armaments and warship works down in Newcastle. When his father died, he inherited the estate and was none too pleased to find that his father had heavily mortgaged the place. From all accounts, Johnny was a good man, and he couldn't stand to see the people who had worked the estate for generations chucked out (McQueen 2008). Pondering what to do, he walked his seemingly bankrupt lands down to the loch. There he fell into conversation with Andrew Lane, a marine biologist and fish farmer, and they hatched a plan to farm oysters. They bought Pacific oyster spats. Pacifics grow faster than the native *Ostrea edulis*. After a couple of mishaps, in 1980 they put a trestle table outside on the road next to the loch and started selling oysters and smoked kippers.

Against all the odds, the enterprise took off. When they started farming oysters, the average consumption of oysters in the entire United Kingdom was five thousand a year. "The company was registered in 1978 with capital of £100. It now has a turnover of £13m, employs well over a hundred people at Cairndow and provides business for its many suppliers" (S J Noble Trust 2014). As the largest supplier in the United Kingdom, they sell the majority of the thirty thousand dozen eaten each year. Noble and Lane were smart, and thanks to Johnny's flair and connections and Andrew's careful tending of the stock, they were soon supplying all the top restaurants in the country as well as far-flung parts of the former empire such as Hong Kong and Singapore (Hoar 2000).

Tragedy struck Loch Fyne when Johnny died in 2002 at the early age of sixty-five. He had done what he set out to do, and had provided for the local people he cared about. He also ensured that money went back into the community for the "relief of poverty, the advancement of education and the protection of the environment particularly within Scotland and the area around the head of Loch Fyne."[7] With the death of Johnny, the future looked grim. Then another player entered the picture. Baxi Partnerships is a trust set up to help employees buy their company. David Erdal, the chairman at the time, helped the employees set up a deal whereby they could buy Loch Fyne Oysters with the help of the Royal Bank of Scotland. The deal worked, and the employee-owners then expanded the business further. It's a story of building on the oyster's capacities for relating, of one community growing and trading in a globalized taste for oysters that has enabled a community to grow, connecting past with present and future. Many of the workers have been there since it opened, and Johnny is still very much present. Some go further back. For instance, the manager of the oyster bar is Christine MacCallum, who is the wife of a farm worker on the Ardkinglas estate, who is himself the son and grandson of estate workers before him. In her words you hear the defiance and pride: "We had been subjected to ridicule for many years as the idea of growing oysters in Loch Fyne was mad and the idea of having an oyster bar anywhere outside of London was, too" (McQueen 2008).

The case of Loch Fyne opens up a number of contrasts. The oyster is at the center, and connections are centrifugally made and remade, be they economic or affective. It is important that it is a foreign oyster. The native oyster, *Ostrea edulis*, has long been in peril, and it was widely regarded as eradicated by the late 1950s. It limps along, and as the property of the

Crown it is strictly protected. It isn't as yet a feasible enterprise. It was the Pacific oyster, the *Crassostrea gigas*, that afforded an oyster industry in Scotland, and although they were hesitant at first to put their feet down, they are there to stay. What happens as the oyster becomes a living commodity? As Yonge predicted in 1960, the oyster has become in some senses a domestic animal, a fond member of community and, in Yonge's words, "a matter of major interest to all, who for one reason or another, care for oysters" (164). It is a particular economy: In most parts of the world, oyster cultivation is an artisanal, familial, or smallholding type of enterprise. Its temporality is governed by the tides and the seasons; the oyster grows relatively slowly compared to terrestrial crops—it takes two or three years to get to market, compared with cereal crops that may have two seasons annually. But the oyster doesn't need feeding or watering or pesticides. It just needs constant care. Oystermen and women are normally hardy individuals who spend a great deal of their time in shallow waters tending the growth of oysters, which proceeds in stages, watching them and turning them in their socks—the open cylinder that is home—until they mature. In flat-bottomed boats they while away their time with the oysters, moving the socks up and down to catch the currents that will fatten them. From my research, they are a happy bunch. As an oysterman in Coffin Bay, Tom, told me, you wait "for the rain that brings the good stuff, and watch for the Westerly changes that bring in food." He also told me he loved his job, though going by the big smile on his face when he talked about his oysters, he didn't have to. Tom radiates oyster love (figure 2.2).

As we've seen, to encounter an oyster in its milieu is to dwell with its culture, to appreciate how oysters intimately relate place and people. Oysters obviously like Loch Fyne. The hardy Pacific gigas took to the low-saline waters where the ocean mix has lost some of its brininess. The Scottish lochs are ancient fingers pushing into the land, which makes the seawater change as it enters the circulation system of the loch. While the weather is often less than kind to humans, it's a fine place to be an oyster. Scientists have recently discovered that oysters are gregarious. They like to put down their feet where other oysters hang out. However, adult oysters are cannibals, and along with starfish and humans pose the greatest threat to young oysters. So wee little larvae cognizant (on some level) of the danger posed by the grown-ups nonetheless prefer to be together. You can hear the excitement of the scientists as they describe how "oyster larvae make a life or death decision when they get their one chance to select where to

Fig. 2.2. Tom Evans (Evans Oysters), Streaky Bay, South Australia. Photograph by author.

attach themselves to the bottom. Our research shows that oyster larvae are willing to risk predation by adult oysters to cash in on the benefits accrued by spending the remainder of their lives among a large number of their species" (Tamburri, Zimmer, and Zimmer 2007).

Oyster lovers, scientists or not, are such that our interest in the bivalve can veer toward anthropomorphism. However, I like to think of this interest as attachment or attunement between researcher and oyster. This goes far beyond the feeling one might have toward an inanimate commodity. And this is not just because they are animate, and animate our appetites and bodies. It is because the ties between consumers, suppliers, environment, history, and the bivalve are so close. This is an artisanal sea ecology, which as Tim Ingold argues arises from "an active and mutually constitutive engagement between organisms and their environments, and in no way precedes that engagement" (1991, 242).

One way to think of this is to turn the economic term "externalities" on its head. In traditional economic theory, externalities are all those costs that producers do not have to—and therefore do not—include in their costings. In his early essay on markets, Michel Callon explains: "Economists invented the notion of externality to denote all the connections, relations and effects, which agents do not take into account in their calculations when entering into a market transaction. If, for example, a chemicals plant pollutes the river into which it pumps its toxic products, it produces a negative externality" (1999, 186).

Callon writes that there are positive externalities as well. He goes well beyond the implied binary of good and bad externalities, and his interest is in the ways in which there are always overflows—of information, interest, money, and so on. These overflows foreground or perform what he calls hybrid forums where actors and their interests are in constant fluctuation. This is a variation on what Latour (2005) calls "attunement." Ben Anderson and John Wylie explain: "An attunement [is] how heterogeneous materialities actuate or emerge from within the assembling of multiple, differential, relations and how the properties and/or capacities of materialities thereafter become effects of that assembling" (2009, 320).

In the case of Loch Fyne Oysters, there are numerous externalities at play, which make us think differently about the relations formed. As I've said, the oyster seems to like the loch. Part of the money provided by the sale of oysters is vested into the Loch Fyne Trust to care for the loch and the now-flourishing small community of humans and oysters in Cairndow. The oyster does its own work of cleansing the loch, and a symbiotic relationship begins. Humans too have had a long, interconnected relationship with the environment oysters call their home. Through the Loch Fyne Trust, which is to say through the sale of the oysters, groups like Here We Are act as an interface between the locals and the incomers, the young and the old. As they say on their website, the objective is "to explain how a rural community, while loyal to its roots, makes its way in the modern world. Its subject matter is people in a place" (Here We Are 2008). I like the forceful statement Here We Are. Thanks to the oyster: Here we stay.

That history is certainly an externality that is tangentially associated with the oyster as commodity. Oyster buyers come from Hong Kong and other places to soak up the history of their oysters' habitat. They are often taken for a tour of Johnny's home, where they are shown his collection of oyster plates, as well as the old-fashioned kitchen and the beautiful scrolled ceilings. This subtly changes the relationship of people to the oyster, morphing it into a form of attachment rather than a pure economic exchange. If we then turn to the story of the laird looking to care for his community, we enter into a scene that economists might more readily recognize as economic. But while he made money from his enterprise, the obituaries and stories about Johnny portray a man not particularly interested in furthering an individual economic gain. He was, of course, motivated by interest, but self-interest or altruism don't quite capture the ways in which his business spread over land, sea, oysters, fish, people, and history.

While we could call this outlook feudal, it is more correctly a profound sense of responsibility and a deep feeling of relatedness, which can be seen as a positive externality—a thick attunement of bodies and histories. Johnny's Loch Fyne produced multiple forms of attachment within the community. This created a kind of cultural and economic ecosystem of oysters and humans, which afforded a form of reassurance. When he died, that ecosystem lived on and allowed his employees to consider owning their own company. In sociological terms, you could call it the conversion of cultural capital into economic capital. As David Erdal, the man who brokered the deal, explains, "Employee ownership gives people more satisfying working lives than they ever thought possible" (Wishart 2008). The power to think beyond "what they ever thought possible" can be included in the long list of positive externalities associated with Loch Fyne Oysters.

Sustaining the Taste for the Sea

Instilling a taste for oysters in this case builds on the interests of place and taste, of human and nonhuman ecosystems. From the lochs of western Scotland, I now briefly turn to another example of how oysters may grow communities, this time in remote southern Australia. Cowell is a small town on the other side of the Spencer Gulf, and down from Whyalla, on the Eyre Peninsula on the wild side of the state of South Australia. This is mining country, which grew out of copper explorations in the 1850s. Mining communities have been swallowed by global multinationals—Whyalla is a one-industry town dependent on One Steel, previously a division of BHP. Iron ore is mined throughout the Eyre, and you can see the tops of mountains that have been lopped off. The red dust covers everything. You turn off the highway and head away from the mine lands toward the sea. Rather quirky fish and oyster art provides a bright welcome into the small town (figure 2.3).

Fig. 2.3. Fish public art in Cowell, South Australia. Photograph by author.

When the ferry is running, Cowell is the first stop on the Eyre Peninsula across from Walleroo. However, the traffic is normally the other way around as people leave toward the east. As with so many small communities built on farming, the town was losing its young people to larger towns or to the mines inland. In 1991, the local oyster growers approached the Cowell Area School (the local high school) with a plan to foster skilled workers in the oyster farming business—to urgently interest young people in oysters as a way of stopping the out-migration. They donated two hectares of oyster leases in Franklin Harbour, a secluded bay rich in sea grass and algae. For a while the aquaculture course worked a treat, and it became a beacon to rural schools in places as far afield as Hawai'i and landlocked Idaho. However, by the beginning of 2009 it was in pieces, with the Department of Education and Child Services threatening to close it down.

When I went to visit the school in 2009, there were signs of new life. The public sculptures that lead into town are fantastical metal arrangements of fish and oysters, and point to the road—Oyster Drive—that takes you into town. A banner outside the school proudly proclaims their revitalized aquaculture program and thanks their sponsors, including Turner Aquaculture and BST (oyster farming equipment specialists, who coincidentally sell to Loch Fyne Oysters). The artwork around the school features boats and oysters and fish and people and the greeny waters of Franklin Harbour and the big blue sky of the Eyre. Jan, the incredibly hardworking principal, led me around the back of the school and introduced me to Mark, an enthusiastic young man who had returned to the school where he did his aquaculture training after having toured the world—including a stint working at Loch Fyne. They showed me a fascinating array of tanks, which they (teachers, parents, pupils, and other volunteers) had painstakingly scrubbed and fixed. Someone had come up with the idea to put windows lower down in the large oyster tanks so that the "littlies" could see what was happening with the oysters. Every school year (or grade) has its own project—from growing the plants of the local marine habitat, to farming sharks and barramundi, to growing the spats, the baby oysters. Alongside this, they all have a hand in the maintenance of the equipment and the leases. And of course there is the hard work of tending to the oysters.

In terms of what we might call the positive externalities of this business, what seems to be emerging is an assemblage of local growers such as Turner Aqua, and connections to Tasmania, where two growers donate spats. This assemblage also includes the buyers in Japan who love buying oysters grown by schoolchildren. The local pubs sell the oysters cooked or beautifully natural. Crucially, the oyster school has become, in Latour's terms, an attunement of community, oysters, and environment. Jan has a solid business plan, which should see that the returns pay Mark's wages. But equally important is the way in which oysters now play a part in the academic curricula in the school, including business studies, food technology, art and culture, and of course marine science. All of these factors are instilling a taste for oysters: The school has become a microcosm of the wider ecology in Franklin Harbour, instilling a desire to learn about taste and place in their area. As we walked among the permaculture gardens fed by the recycled hydroponic nursery, I still was curious. "But do the young kids like the taste of oysters?" Mark laughed and said that when they were shucking, the kids popped oysters in their mouths as if they were sweeties, or candies.

Here's some of the things that a taste for oysters can do: young kids slurping up the oysters that grow with their school, that they've helped to take care of—from the nitty gritty of cleaning out the oyster socks, or sluicing off the boat, or helping to grade them. They look up to Mark as a worldly figure (still in his twenties) who roamed global oyster-growing spots until, like Ulysses, he returned home to tend his oyster patch. This is a multigenerational, multispecies assemblage: Sons and daughters follow their parents into the oyster life; the relatively recent newcomer Pacifics have afforded the return of the ancient angasis. Children here relish something other Western kids would dismiss as "sea snot." Taste is both literal and figurative. Children acquire a taste for something that is central to their small community, and that has animated and related a renewed sense of belonging to the fertile basin of Franklin Harbour.

Then there's the local superstar of oyster growing: Brendan Guidera over at Pristine Oysters. The winner of numerous awards and fellowships that have taken him to places like Loch Fyne to gather and exchange knowledge about beloved bivalves, Guidera produces about six million Pacific oysters a year. He has now re-created the environment to grow the native *Ostrea angasi*, the oysters that were dredged to extinction in the late nineteenth century. They are almost identical to the native oyster, *Ostrea edulis*, found in Scotland and in Europe. These are the ones that the French love, and Guidera has eager Parisian buyers competing with the chefs at the top restaurants in Sydney and Melbourne. He reflects on his life with oysters: "As we clear off the barnacles, salmon and trevally swim around our legs," he says. "We bring out an Esky, a frypan. It's a crackin' lifestyle" (Powell 2009).

Nach Urramach an Cuan (How Worthy of Honor Is the Sea)

Across this chapter I have tried to convey the temporality, the ebb and flow of humans and oysters. The affairs of oysters wax and wane—they are dredged into oblivion and then somehow come back again. But still I was shocked to learn in 2008 that the owner-employees had sold off the Loch Fyne Restaurants arm of the business to Greene King, one of the largest hotel and pub owners in the United Kingdom. The forty-two restaurants (all down south in England, bar the one on the Firth of Forth) became themed as Loch Fyne Seafood and Grill. Does it matter that they are an English megacompany with a turnover of more than three billion pounds in 2014?[8] They continue to use Johnny Nobel's story and have made the Gallic saying *Nach Urramach an Cuan* into their philosophy. In 2012, I returned

Fig. 2.4. Loch Fyne Oysters kippers smoking at the Inveraray Highland Games. Photograph by author.

to Loch Fyne to follow up on what had happened. I was told to meet the public relations people at a local Highland Games show over in Inveraray. The town was jumping, and I ducked the brawny guys getting ready for the caber throw and the young girls dressed in Highland dancing outfits, their mothers fussing over their plaits and makeup. Mingling throughout was the fugg of the Loch Fyne kippers being smoked (figure 2.4). But where were the oysters?

While I love the smell and taste of kippers, and those from Loch Fyne are very good, things felt a bit off. Just earlier that year, the business was the subject of a takeover for a rumored seven-figure sum—millions of pounds—by a coventure of Scottish Seafood Investments, the Scottish Salmon Company (formerly known as PanFish), and a private equity investor, Northern Link Ltd. A news site, ForArgyll.com, reported, "The former employee-owners of the business will get shares in the Scottish Salmon Company but will not be paid a dividend as part of the sale" (ForArgyll.com

2012). The company had been struggling, and one can hardly blame the employees (no longer employee-owners) for wanting a more secure future. Inevitably, the story was reported in an uplifting way as being about "Scottish enterprise." However, this didn't deter several Scottish respondents. Commenter Tarbert Saxon wrote, "I fail to see how this is good news. A locally owned, locally managed, local employer taken over by a company owned by a Scandinavian company that has had more name changes than Windscale and a private equity company" (ForArgyll.com 2012). Another writer, Ewan Kennedy, replied, "They appear to have two executive directors, Robert M. Brown III, ex–Lehman Brothers managing director and resident in Moscow, who is involved with Kazakhstan and Russian investments and Vlacheslav Laventyev, who has a long involvement with Russian fish farms." Kennedy ended his post with this remark: "There are no doubt good reasons for the sale but I find the decision rather sad, as successful employee-owned businesses add a lot to their communities."

Tae Think Again of Loch Fyne

It doesn't sound good, and in late autumn 2014 I return to Loch Fyne. But as I sit by the loch in Inveraray on an unusually fine day, the picture doesn't look so bleak. The small town is heaving with tourists as the tour buses disgorge people from around the world. I hear broad Québecois alongside Mandarin in the Loch Fyne Whisky shop, as people buy up the bottles of fine single malt.

Later as I sit at the beautiful new bar at Loch Fyne Oysters and contemplate my mixed half dozen oysters, things seem pretty shiny. I am there to meet with Penny and Matt from the Scottish Salmon Company, which had gobbled up the little enterprise. They are not quite the villains I was led to expect. Penny considers herself a conservationist first and foremost. She left a job in whale conservation to join the Scottish Salmon Company. She speaks with a soft Highlands accent about her passion for the area and for the possibilities of aquaculture as "the answer for producing quality protein with certainty." There are technical problems and cultural or social ones yet to be fully resolved. On the technical side, she is excited about the possibilities of integrated multitrophic aquaculture, which could see fish being farmed in conjunction with oysters, mussels, algae, and even sea cucumber. Like Penny, I'm fascinated by the potential of this system to produce truly sustainable marine-based protein, and I return to it in chapter 5. For Penny and Matt, the current task is to educate local people about how the industry of fish farming is changing. As they put it, it is a young industry, and over the space of one generation people have gone from seeing the excesses of aquaculture ("They'd chuck bags of fish feed straight into the loch") to the modern technological setup it is today.

Later I'm lucky enough to be taken on a tour of the Loch Fyne Oyster sheds with Virginia Sumsion, the niece of Johnny Noble. She joined her uncle in 1990 as a sort of event manager, although it is readily apparent that Virginia turns her hand to all sorts of tasks. She is above all "evangelical about oysters." She describes her job as a vocation "to get people to eat oysters." She shows me the high-tech equipment that takes the oysters from the loch to tanks that have loch water running through them. The addition of UV light kills any bacteria. It's a big modern business built on the backs of the oyster and her uncle's passion. From the shed they are shipped around the world. We are joined by a suave businessman from Hong Kong. He buys most of his oysters from Loch Fyne, and he has always wanted to visit.

We drive over to the Arkinglas House, a grand old house, which was Johnny's parents' summer home. There Virginia expertly shucks the oysters she picked up in the swirling container pools in the sheds. She's brought flat oysters, the native *Ostrea edulis*. It is a very different taste and texture than the Pacifics I am so accustomed to. Very briny, and they fill my mouth with sensation. Strangely enough, the buyer is allergic to oysters, even though he still loves them. We eat his share.

In response to my questions about whether they have felt a loss with having to sell the employee-owned company, you can see the relief on her face. She tells of the ups and downs that they have had ever since Johnny and Andy Lane had their brilliant idea to grow oysters. The influx of cash into the company has put them on a firmer footing. Perhaps the extraordinary thing about the whole venture is that in an isolated part of an isolated country, there is now no unemployment. People commute from thirty miles away, and more houses are constantly being built on the estate. At the very beginning of the plan for Loch Fyne Oysters, Andy Lane and Johnny Noble agreed that the enterprise was to provide "real jobs for real people."

The tale of oysters and humans continues—in places as far apart as Cowell and Loch Fyne. It is a tale of contrasts: new oysters and old money in Scotland, high school children tending oysters in the most southerly part of Australia's mainland. Very different people are related through oysters, and the oysters have allowed two different small communities to become quietly and moderately prosperous. In Argyll and Bute, growing oysters has enabled the We Are Here community center to collect oral histories that fill in the blanks left by the Clearances. In South Australia, the onus is on bringing up children to know and care for the marine environment in which their oysters grow. Through their small enterprise, school kids now have horizons that would have seemed impossible for their parents, and connections with a wider world.

In their article "Mixing Methods, Tasting Fingers," Annemarie Mol and her team contemplate how "the varied literatures on eating practices have little or nothing to say about taste" (Mann et al. 2011, 222). Their experiment involved different challenges than mine. Theirs is about what happens to taste when you communally and reflexively eat with your fingers. Mine has tried to make taste cover a broad spectrum, to locate oyster eating as a centrifugal moment that throws out connections to history, class, place, and mouths. The challenge they put to themselves reverberates: "How to foster research practices that afford rich and layered realities? How to make space for analysis as well as care? And how, finally, to keep complexities, ambiguities, tensions and pleasures alive in one's writing?" (Mann et al. 2011, 229).

At the heart of my tale is a desire to relate the materiality of oysters and of their singular ability to extend the human imagination. In this sense, I follow the oyster as it relates to a more-than-human ecology. I also relate the stories fed by the taste and feel of the oyster, of its capacity to clean water, to maintain its own milieu alongside that of humans. This isn't a one-sided affair: While over history oysters have been dredged to near extinction, swamped with human pollution, and eaten without thought for the fate of species, some human-oyster relations bear the stamp of the *interesse*; of a being together in the in-between. This is a fine thread that continues in some places where people may learn about the loves and lives, the sex and gender of the oyster. Like Virginia Sumsion, the ambassador for Loch Fyne Oysters, I really want to encourage others to experience the zincky slide of a Pacific or an angasi down the throat.

In chapter 3 I return to South Australia, this time to swim with one of the most magnificent fish in the world—southern bluefin tuna. Their tale is a very different one, where the outcomes are much more difficult to map. I also want to ramp up the stakes of what it means to eat the ocean.

3

Swimming with Tuna

I'm going to swim with tuna. I'm going to swim with bluefin tuna. I'm going to swim with large endangered fish.

In chapter 2, I followed oysters as they compelled new ways of understanding more-than-human relatedness. Following Mol's (2008) cue, I experimented with the refrain "I eat an oyster." Focusing in and through the oyster allowed me to explore new dimensions of taste, tastes, and tasting. Tasting an oyster is a mindful, or "care-full" (Miele and Evans 2010) experience, which as I argue is a necessary ethical practice if we are going to attempt to eat the ocean well. It alerts the eater to the complexities of oceanic interrelations and helps to place the human as but one part of who eats whom in oceanic relations. In this chapter, I want to push the boat out further.

To say "I eat bluefin tuna" would immediately put me beyond the pale. One of my interviewees, Caroline Bennett, set up the first fully certified sustainable sushi restaurant in London. She recounts an exchange she had with her fish seller several years ago, when she wondered why she could no longer easily access bluefin tuna. The fish buyer responded, "Would you put white rhino on your menu?"

The World Wildlife Foundation's very clever advertising campaign has an image of a tuna with a panda mask, asking, "Would you care more if I was a panda?" (Ads of the World 2014b). The campaign concocted by Ogilvy Paris also uses images of tuna with rhino masks, again imploring, "Would you care more if I was a rhino?" While slick and smart, this campaign does not hit quite as viscerally as does the one for Sea Shepherd Conservation Society's Blue Rage. Made by Ogilvy Mather Singapore, the series of images features realistic-looking shots of a fishing boat, complete with bare-torsoed fishermen, above whom hang the mutilated corpses of pandas. In another one, in what looks like Tokyo's Tsukiji Fish Market, buyers walk along the lines of what would be frozen tuna, but here again they are replaced with bloody open-eyed dead pandas. The caption for the campaign is "When you think tuna, think panda." It goes on to explain, "The bluefin tuna is now critically endangered to the point of extinction. Industrial overfishing, fueled by the voracious appetite for tuna in Asia, is killing off all the breeding populations" (Ads of the World 2014a).

Of the two campaigns, the Sea Shepherd's Blue Rage works most blatantly to incite disgust. The image of pandas hanging by their feet instead of tuna by their tails, and the realistic rendering of their bloody bodies on the ship's deck, hits forcefully. I can feel the taste of bloody matted fur in my mouth as it juxtaposes with memories of smooth tuna flesh. The combination makes me gag. The somewhat cute campaign with tuna decked in panda or white rhino masks just doesn't grab me in the same way. In another mode of campaigning, the FishLove site I've described earlier also doesn't reach the guts. The sight of Helena Bonham-Carter lounging nude on a bluefin tuna or Jerry Hall cuddling a tuna in her bare breasts doesn't bring forth disgust; maybe bemusement, or the interest of interspecies erotica. When I ask Caroline Bennett what she thinks, she's dismissive: "It's slick with no statement."

"I swim with tuna"? Might this be a way into an appreciation of the many facets of human and tuna relations? What forms of relatedness might we find in the ocean together? There's something magically both in and out of place about human bodies in the ocean. As Steve Mentz argues, "To swim requires giving oneself over to the alien element" (2012, 589). In his book *At the Bottom of Shakespeare's Ocean*—a monumental undertaking to read seemingly all of Shakespeare through the optic of a "blue cultural studies"—Mentz writes, "The sea touches human bodies most intimately through the halting and laborious art of swimming. We know we're in it— or at least we feel like we know it" (2009a, 35). He elaborates on how "depictions of swimming expose the awkward fit between human bodies and the ocean" (35).

I take this point to heart. I used to think that I was in my element in water (despite being born under the air sign of the water bearer). One of my favorite pastimes used to be swimming straight out into the middle of the ocean—often at night. One time I was bewitched by the bioluminescence of the sea. Swimming back I encountered flock after flock of bluebottle jellyfish, and I emerged on the shore flagellated. My habit has been curtailed by my partner's worries about sharks. Australia is known to have the world's highest percentage of fatal shark attacks. Her concern ("How would I tell your father you were killed by a shark?") has made me slightly more prudent. In her autoethnographic account of becoming a marathon ocean swimmer, Karen Throsby writes of the moment when she thought she saw a shark: "A fear-filled rush of adrenalin; my heart pounds, my seasick stomach knots coldly, and the backs of my hands, head and neck prick sharply, like the skin is lifting up from the muscle and bone" (2013, 6). Throsby nonetheless takes the plunge: "Being *in* is always better than being on the water" (6).

I tend to swim on the water rather than in the water. To my chagrin, I have too often floated above the densely populated world of the sea— analogous to a critique leveled by Aboriginal people about how "white fellas" walk through country rather than in it. It's a one-sided view of the world that I need to redress if I'm to do some small justice to the immensity of how to live with the ocean better. It is also good to remember that, as Mentz says, the fit between human bodies and the ocean—as well as her myriad inhabitants—is an awkward one. Swimming doesn't produce easy reconciliation between and among all these elements; in fact, it may act as a necessary practice of estrangement that summons the awe that humans should have before and in the ocean.

To set the scene, the reason why I've come to Port Lincoln, South Australia, to swim with tuna is because I can. Port Lincoln is the mecca of tuna fishing, ranching, or farming in the Southern Hemisphere. And you can swim with them. When you arrive in town and drive down toward the shore, far out to the horizon of Boston Bay there are over 150 tuna farms hidden deep under the water. Thousands of southern bluefin tuna (*Thunnus maccoyii*) are being fattened for slaughter. It's a very different way of rearing food than the simple life of the oyster up the road in Cowell. And as I will find out, the men who farm tuna are of a different breed than the oystermen and women in chapter 2.

There are two adventure outfits that offer tours where you swim with tuna. From what I can find, the only other place in the world that offers tuna-swimming tourism is Malta, another bluefin tuna farming hotspot. On the advice of a friend, we decide to go with Adventure Bay Charters. The owner, Matt Waller, got out of tuna fishing and came up with the idea of a fully eco-offset enterprise. He's still only in his thirties (they start early on the boats) and is immensely hardworking. Tourists from around the world come to encounter tuna and sea lions at first hand. He has recently added swimming with great white sharks, which strikes me as weird. You don't really swim; wet-suited humans sit in cages waiting for the sharks. The other option of swimming with seals allows you to snorkel with what are called "the puppy dogs of the sea." The local fishers call them "the great fish extractors" of the sea, and claim their populations have exploded to the detriment of fishing.

Matt is clear-eyed about the difficulties facing the fishing industry, which include the fierce and ultimately ecologically untenable competition for dwindling worldwide stocks. He points to the long hours and the still very dangerous nature of the work. But he also remembers vividly what he calls the "high of fishing." Those were the days when the fish were easy and Port Lincoln would be full of young, hardworking fishing guys with lots of money and ready to party. Money doesn't flow quite as easily these days, and the town is besieged with worries about the future of the fisheries and of fishing itself. Matt himself is in a fierce price war with another outfit that coined the name Swim with the Tuna some years after Matt had initiated the idea. Such are the emotions that normally voluble fishers around town go quiet when I mention the feud. Whereas Matt runs with a smallish boat and pontoon, the other company has a bar and café out at sea. It looks like a miniature Sea World with tuna pens. Later I learn that Matt lost out to the other guy. It's not only fishing that is a hard game.

We board the zephyr down at the marina, are fitted with life jackets, and then zoom off. Soon we are in the middle of Boston Bay, which is a nook at the bottom of the Spencer Gulf. All around us are huge pens. From the air, they look like black circles in the clear blue water. The ride over in the tour boat is enough to churn my stomach, and the sideways swaying of the pontoon isn't helping my general queasiness. Steve, our handsome tour guide, has given us all sorts of information gathered over some twenty years of his life as a prawn trawlerman, deep sea fisherman, and now occasional guide on his mate's tour boat. The floating enclosure is heavily protected against the pirates who steal the valuable fish. In front of me is a pen containing some three hundred juvenile southern bluefin tuna. The idea now is to jump in with them as they speed for pilchards—cousins of sardines. They can go from zero to fifty miles per hour faster than a Porsche. I just hope that they brake well.

Paul Greenberg's description captures the sheer awe-inspiring fact of bluefin tuna: "For those of us who have seen their oversize-football silhouettes arrive, stop on a dime, and then disappear in less than a blink of an eye; for those of us who have held them alive, their smooth hard-shell skins barely containing the surging muscle power within, they are something different than the space they occupy" (2010, 200).

The ones in our pen are young, but nonetheless they seem pretty big to me. Steve and his young helper show us how the tuna rise to the surface when they are fed. He has several crates of pilchards. He shows us how to carefully hold the fish by its tail above the water so that the tuna don't get your hand in their haste for food. Commercial pens hold a couple thousand tuna, and they consume huge amounts of feed—whole blocks of frozen fish are winched into the pens. Even relatively small operators will spend three million dollars on pilchard feed a year. Unlike salmon farms, the tuna farms in the Eyre Peninsula apparently don't use antibiotics. Badly managed salmon farms have been responsible for contaminating wild stocks both genetically and with disease, and contributing to the pollution of oceans. Because tuna pens are further out to sea where the tides disperse the debris, this is less of an immediate problem. Nonetheless accidents happen. The tuna farmers of Port Lincoln and the town's inhabitants vividly remember a catastrophe in the mid-1980s when a huge storm caused the ocean currents to quickly change, and all the farmed tuna died in a few short hours. Some seventy thousand of the tuna drowned as their gills filled with silt. Fish drowning—it's a truly dreadful image. A huge amount of money was lost, dead tuna floated in the water, and the stench of lost life covered the town.

Fig. 3.1. Southern bluefin tuna, Port Lincoln, South Australia. Photograph by author.

It was a catastrophe for the tuna and the humans who owned them—the latter lost fifty-five million dollars.

No one mentions that today. There's a lot of milling about and chatting on board. People slowly drag on their wet suits. You can see the fish under the surface of the water, and I can't wait any longer—and in any case I hate the feeling of a clammy wet suit isolating me from the water. I fit my goggles, smile to my partner, who wouldn't do this to save herself, adjust my swimsuit bottom. And I dive in. It's a weird feeling diving into a pen filled with fish. I can feel them beneath my feet. They are so gloriously sleek. I have the grace of a blow-up doll bobbing on the surface above them. They tickle my feet. Then the young boy helping out on the tour chucks pilchards at me. Sure enough, these sleek beauties head toward me at top notch, get the fish, and then stop just short of my nose (figure 3.1).

Floating above the bottom of the sea, dodging pilchard debris, and commingling with tuna—it's a rather queer introduction to bluefin tuna. Swimming with tuna provokes many ethical questions, to which I return later. But I also want to use swimming as a methodology of encounter, albeit an athwart one. To recall, for Helmreich the term captures the need to work "the empirical transversely" (2011, 134). This aspect of athwartness is important as I engage with the different scales of tuna: as global commodity, as endangered species, as the basis of diverse communities. But Sedgwick's (1993b) description of athwart captures how I swim with the tuna throughout this chapter: caught in tides of "continuing movement, motive—recurrent, eddying, *troublant.*" Caught within strange currents, I am swimming athwart trying to figure different ways of understanding this beautiful fish.

On paper at least, swimming with tuna follows David Goodman's take on the enmeshment of social and natural: "to explore nature-society co-productions" (1999, 34). While I may have learned a fair amount about tuna through reading about the various aspects of their life—from the analyses of fishing trade figures through to ideas about what makes their flesh so delicious—being in their milieu (or at least in their medium) brings to the fore other facets. Wallowing in their pens where humans seem to be at the beck and call of the tuna ("feed me"), I am viscerally out of my domain. Indeed I feel very green. I am once again feeling seasick—physically and intellectually in Pálsson's (1994, 905) sense: viscerally lacking in enskilment. Here ontologies are unmoored. Bluefin tuna, as we will see, complicate any strict division between wild and domestic, natural and private property. Fiercely wild fish, they are now raised in marine feedlots. But tuna exert a powerfull pull on humans, inciting them to migrate across the world in their search for these magnetic fish.

The Making of Global Tuna: From Nature to Commodity

Bluefin tuna are the emblem of globalization. Many other commodities are routinely given this title (oil, McDonald's, etc.), but tuna are global in every way. They swim across the globe, and, now orchestrated by the complex trading headquartered at Tsukiji, fish become global commodity.

Southern bluefin tuna spawn in the Indian Ocean near Java, and from there the juveniles make the trip down and around the western coast of Australia and into the Great Australian Bight, a huge open bay that stretches over a thousand miles. As figure 3.2 shows, the tuna regularly travel from the bottom of South America, past Africa to Australia, before heading around back to South America via New Zealand. It goes without saying that they are amazing swimmers. They have to keep moving because they get oxygen from water flowing across their gills. In their long journeys they travel at

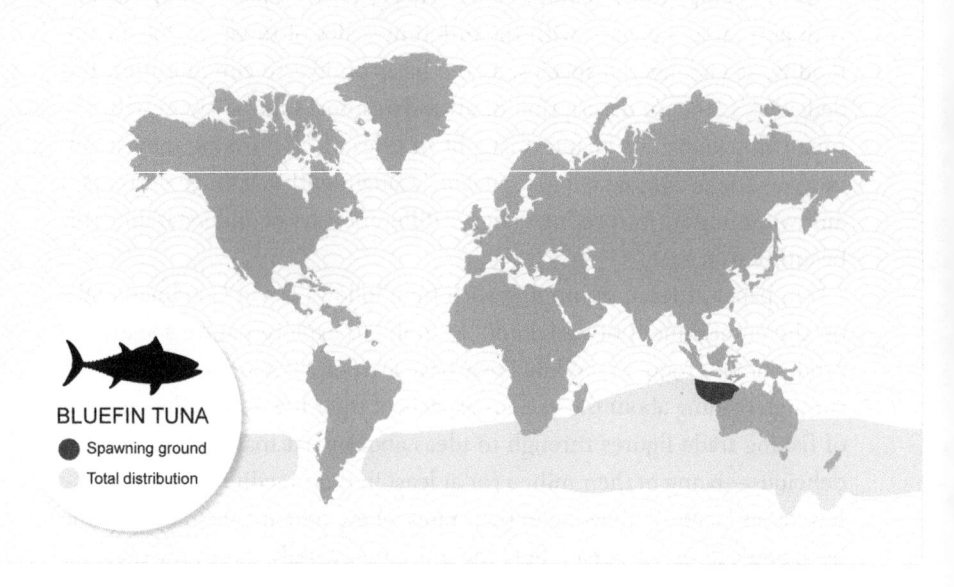

Fig. 3.2. Map of southern bluefin tuna migration. Illustration by Morgan Richards.

up to fifty miles per hour. Their fins retract into their bodies to make them into the most perfect aqua-dynamic eating-swimming-breathing machines. For me and for many, this is what makes tuna so special. For fisheries management, this is why they pose a particular problem in that they routinely range across the jurisdictions of so many countries. As a highly migratory species, they transcend any single state's jurisdiction, which means that multiple states can claim an interest in resources inside of individual states' Exclusive Economic Zones (Campling and Havice 2014). As Robin Allen of the FAO argues, "Free riding states would be able to enjoy the benefits of the efforts of conservation made by responsible states" (2010, 2).

Humans have recorded in various ways their millennial associations with bluefin tuna. In Australia, coastal Aboriginal people would have caught tuna using baskets when they came into shallow waters. The Aboriginal artist Anchor Kalunba created beautiful baskets inspired by the fish traps of his country. While Kalunba is from Kuninjku, Arnhem Land, which is a long way from the Eyre, his *mandjabu* (barramundi fish traps) look very much like a prototype of the purse seine nets now used commercially.

The techniques and technologies that mediate human-tuna interaction have changed considerably, leading to huge increases in tuna's value and of course massively decreasing their stock. As I mentioned, the greatest seafood market in the world, the Tsukiji Fish Market in Tokyo, sets bluefin tuna's global worth. An Indian journalist, Abhijit Dutta, breathlessly describes his first impressions: "It could have been a high temple, a grand mosque or a magnificent church, but it is, in fact, the world's largest fish market. Every morning, from roughly 3–11am, 60,000 people go to work at the market, a space the size of 43 football fields, and get busy with cleaning, cutting, carrying, auctioning, and selling three million kilograms of fish" (Dutta 2015). Ninety-nine percent of Australia's bluefin goes directly there—sometimes fresh, which attracts high value; however, mostly it's flash frozen at minus seventy-six degrees Fahrenheit. The first auction of the year is always important. There are the kudos of buying the first big tuna, and at New Year celebrations people are ready to pay top dollar for the most prized part of the tuna, the *toro* or belly. In 2013, one man (the owner of a large chain of sushi restaurants in Tokyo) paid 1.76 million dollars for the first tuna of the season. It was big—nearly five hundred pounds—but the price was seen as part showmanship on the part of the buyer. Nonetheless it is staggering. How does a fish come to cost that much?

The story of bluefin tuna's rise in value is one of technology and taste. In the early twentieth century they were called "horse mackerel" and were seen as a pest for fishers because they would follow smaller fish, eat them, and then tear through the fishing nets. Off the East Coast of the United States and Canada, Atlantic bluefin were common and their size and strength made them the darlings of sport fishermen. The practice was to catch them and then back at the dock get a picture taken with a gleaming giant tuna. The fish would then either be thrown on the town dump or chopped up for cat food. This remained the fate of tuna until the 1970s, and even with the most zealous of sport fishermen bluefin stocks were healthy. But all of that was to change. The story goes that a Japanese businessman witnessed what must have to him been a bizarre ritual of catching, photographing, and dumping a beautiful fish. He may have had an inkling of the money that the worldwide sushi business was to generate.

In the 1970s, Japan Airlines began to experiment with loading tuna from the East Coast of North America to fill their empty cargo holds after delivering electronic goods to American markets. As Dutta (2015) writes, "Desperate to make the return flights commercially viable, a young JAL cargo manager in Tokyo, Akira Okazaki, bet on tuna—he needed something that would demand a high price from Japanese buyers and was perishable enough to merit the high cost of air transport." The anthropologist Theodore Bestor (2004) provides one of the most in-depth and wide-ranging accounts of bluefin through the lens of Tsukiji, which he calls "the Fish Market at the Center of the World." Bestor describes the work of the "middlemen" of the tuna trade as "the technicians of globalization" (2001, 77). This account describes the role of bluefin that propel and are caught in "flows of capital, both financial and symbolic, in multiple directions." What stands out is the way in which seemingly peripheral places like New England, the Canadian maritime provinces, and, as we shall see, Port Lincoln become connected to the high-finance global empire of the Japanese tuna industry, and along the way also become interconnected with each other. Bestor (2001) describes this as the globalization of a regional industry.

As bluefin becomes a marker of cultural and economic capital, it also changes people's tastes around the world. This happened in a remarkably short amount of time. Sasha Issenberg describes how tuna went from being pet food to one of the most valued natural resources in the space of two decades as an event "that has little parallel in history" (2007, xii). Intriguingly, the Japanese formerly didn't like the taste of bluefin, thinking it too bloody and fatty to be eaten raw. However, apparently they became accustomed to the heavy taste of beef during the American occupation following World War II. Feeding the starving Japanese population was seen as a crucial way of persuading them of the merits of democracy. One of General MacArthur's first actions was to set up a food distribution network. Japan has always held its fisheries in high esteem. As an island nation with little arable land, for food security it has always had to look to the seas. Indeed, as Kate Barclay and Charlotte Epstein (2013) argue, the fisheries were seen as key to the successful industrialization of the country.[1] So while the Americans may have accustomed the Japanese to the taste of fatty, bloody meat, their fishermen were out trying to catch as many tuna as they could. This became much more intense in 1952, when the prohibition against Japanese boats fishing beyond their territorial limits was lifted. Not only did they need to feed their battered nation, Japan also needed hard currency to rebuild. Another theory is that "the fatty belly-meat of bluefin tuna called otoro and shimofuri . . . became a delicacy as a result of a Japanese government marketing campaign to deal with tuna shortages caused, it's said, by nuclear weapons tests in the Pacific" (Renton 2006).

How America came to have a taste for tuna is also a fascinating tale. In some accounts the Kawafuku Restaurant in Los Angeles was the first to serve sushi in 1962. But it was the invention of the California roll that really popularized sushi in the United States and then the world. As John Mariani writes, this was "a form of sushi made with avocados, crabmeat, cucumbers and other ingredients wrapped in vinegared rice. It was supposedly created at a Japanese restaurant in Los Angeles named Tokyo Kaikan about 1973 for the American palate but has also gained popularity in Japan, where it is called kashu-maki, a literal translation of 'California roll'" (1999, 53). Obviously crab or fake crab should have had a minimal impact on tuna fisheries. However, as Greenberg astutely notes, "the West's embrace of the Japanese sushi tradition had a multiplier effect: it brought people who had previously disliked fish into the fish-eating fold" (2010, 203). Greenberg gives a convincing reason why people who don't like fish (and in my experience there are a lot of them) like raw fish. Tuna and especially bluefin tuna swim hard and have to use and store large amounts of a chemical called adenosine triphosphate. In raw tuna this converts to another chemical called inosine monophosphate. This produces the umami tastiness, the fifth sense.

Fishers Following Tuna

From the moment when a Japanese cargo technician thought to recoup costs of air transport to American markets by shipping bluefin tuna back to Tokyo, to its current proliferation around the world in high-end sushi restaurants, bluefin becomes the emblem of globalized trades and tastes. The seemingly inevitable outcome of this ramping up in value and in global taste is that the stocks of bluefin are now seriously depleted.

I want to turn to the story of how fishers in South Australia pioneered a response to this worldwide depletion. Their idea to farm bluefin would radically denature tuna. This is a determined tale of how a supremely wild fish becomes a domesticated animal. Framed as such it is tragic. But as we return to South Australia to meet some of the key players in this domestication, we also encounter a love story of how fishers fled their homelands to follow the scent of tuna to the other side of the world. This too is part of the global tale of tuna.

Not long after I swam with tuna, I met the men who farm them.[2] Modern Port Lincoln was made through tuna, and a handful of men made the tuna into machines that produce huge amounts of money. Port Lincoln is widely known as the capital of Australian millionaires, money made exclusively from tuna. The town is littered with their toys—or replicas thereof,

Fig. 3.3. Dinko Lukin, Port Lincoln, South Australia. Photograph by author.

in the case of the racehorse Makybe Diva. Her statue stands proudly on the green foreshore. She is owned by Tony Šantić, who named her by taking the first two letters of the names of his five employees—Maureen, Kylie, Belinda, Diane, and Vanessa. Makybe Diva won the Melbourne Cup three times, and by the time she retired she had made her owner fourteen million dollars. Šantić, like all the tuna families, is Croatian. The tuna trade had been split ethnically, with the Italians running the east coast of Australia and the Croatians—all from the little island of Kali—taking South Australia. The famous Puglisi family straddled the two, and Joe and Mick Puglisi sailed from the east to cash in on the tuna trade in 1968. Joe sold his quota for 180 million dollars in the 1980s. Now all the tuna action is in Port Lincoln and, with the exception of Hagen Stehr, it is all in the hands of the Croats. They are a flamboyant crew. Šantić owns Tony's Tuna International. The logo for his company is a cartoon of him as a long-haired dude with sunglasses sitting in a boat with a fish on the hook.

One of the tuna barons—in his eyes, the true baron—is Dinko Lukin. In 2010 I meet Dinko at his factory just outside of town, which certainly doesn't look like a millionaire's pad (figure 3.3). Dinko is now in his late

Fig. 3.4. Dinko tuna pens, Port Lincoln, South Australia. Photograph by author.

seventies, and he arrived in Melbourne from Croatia in the 1950s. I am early, so I wander down to the water to look at the resting farming pens. These are the commercial ones, much bigger than the one I swam in. Pelicans and other seabirds line the pen, waiting hopefully for some action (figure 3.4).

Back at his office, I meet Dinko and a young woman—both have stunning blue eyes. It turns out that the young woman is Lukina, his Thai second wife, and it seems that she wears contact lenses to match Dinko's eyes. They met at a Thai restaurant in Port Lincoln in 1996, three years after Dinko's expensive divorce from his first wife (a common occurrence in high-value fishing circles—and indeed gossip has it that one fisher keeps on going out to sea long after he wanted to retire to avoid the wife and inevitable divorce proceedings). Lakanna was twenty-nine when they met and Dinko was sixty-one. To show her devotion she changed her name to Lukina, which means "belonging to Lukin" in Croatian. We sit in his crowded office, and he looks at me intently as he tells me some of his life story.

SWIMMING WITH TUNA 89

Soon it turns out that he is a bit of a charmer, and he has me girlishly giggling. I am to find out that the tuna boys are a seductive lot, something to do with those eyes that seem to reflect the ocean—and how they exude intensity, money, and power.

In the 1960s, Dinko commissioned his first tuna boat, the *Orao*, which held a crew of seven. This was in the days when tuna were long-lined and poled. The Japanese had been long-lining in Java, where the tuna spawn, and in Australian waters since the 1950s. This requires large boats, which subsequently became the infamous factory ships when the Japanese came up with the technology to freeze fish on board at minus sixty degrees. In the early 1960s, the Japanese boats would catch over 81,000 tons of southern bluefin tuna a year. By comparison, in 2014 Australia was allowed to catch 5,000 tons.

Back then, Dinko's little boat was limited to surface catches with lines and poles. The tuna were caught on lines and then hoisted up by pole. This is the method that is now deemed more sustainable, and it's this that will be mentioned on your tins of "responsibly fished" albacore or skipjack tuna. It seems strange to think that it was a couple of American brothers who taught the Australians this method, but then as I said this is a very global enterprise. The Jangaard brothers introduced the Port Lincoln fleet to live bait fishing—tossing or chumming with live pilchards keeps the tuna school in a feeding frenzy (ABC 2013). It didn't do much for the tuna, who were bruised in the operation. Given their current monetary value, that's unthinkable now. The work was hard on the men too—you can imagine what they faced in the waters of the Great Bight. For instance, Joe Puglisi's story was common: "Conditions onboard were hard. Before it was lost, Joe's home for seven years was a cramped bunk on the St Michael. He went to sea at 13, he slept on a horse-hair mattress, his worldly possessions fitted into a single drawer. And when it rained, he wore his waterproofs to bed." Looking back on it, Joe remarks, "These blokes had balls, I'm telling you. They were bloody. . . . They had to be, just to survive" (ABC 2013).

Dinko's son, Dean, was a testament to the muscles required—he was the first, and so far the only, Australian to win an Olympic gold medal in weight lifting in the 1984 Olympics. He would test his talent in the famous Tuna Toss at the Tunarama, a festival of all things tuna hosted by the now-closed John West cannery in Port Lincoln. As Dinko describes it, life was pretty good in those early days. "The crew did not feel any fear as they faced the strongest elements that a god created." Not only didn't they fear the elements ("I'd never give one inch to the weather"), they preferred going out in bad weather. It was said by his crew that the *Orao* was a magic boat, often hauling in a huge amount of tuna when no one else could. Later magic was aided by technology with the use of spotter planes, which fly low over the seas, reporting down to the boats on the location of the large schools that swim close to the surface. Some spotter pilots brag they can tell the size of the fish to the nearest kilo. The tuna don't have much chance.

With the downturn of stocks in the 1980s, the fish seemed to disappear, along with assorted species of other fish caught up in the chase for tuna. Long-lining is seen as responsible for the worst of bycatch. Now mainly used by the Japanese (according to non-Japanese sources), these boats work twenty-four hours a day laying out fishing line up to one hundred kilometers long, each carrying up to three thousand baited hooks. Blaming long-liners and the Japanese for the downfall in stocks doesn't bear much scrutiny. As I discuss later, the Japanese taste for raw tuna has spread around the world, and Japan remains the hub for world prices on sashimi-grade bluefin tuna. Less reported is the considerable debate about fishing methods within Japan. The Tokyo-based Organization for the Promotion of Responsible Tuna Fisheries (OPRT) was established to reduce the large tuna long-liners from around the world. According to *World Fishing and Aquaculture*, a U.K.-based journal for commercial fishing, OPRT has managed to bring on board 90 percent of the world's large-scale tuna long-liners over twenty-four meters in length, in twelve countries. One of the significant problems in steering the global industry toward more sustainable fishing is the number of boats that sail under flags of convenience—which is to say, those that take on flags from smaller and nonmember states who routinely disregard national and international attempts to regulate the industry. Japan for its part has committed to not buying fish from such vessels; however, the routes of fish sales are circuitous ("Japan Urges Industry-Wide Reduction" 2010).[3]

Bluefin tuna is a highly political fish. When its monetary value was realized, and concomitantly stocks plummeted, the tuna nations had to wake up to the fact that if they didn't do something, they would lose the cash cow altogether. Between 1960 and 1980 the global catch of bluefin had halved. In 1984, the Australian government decided to implement the system of individual transferable quotas (ITQs). Based on their year's catch, fishers were given a percentage of the overall total allowable catch. As the name indicates, ITQs are awarded to individuals and can be sold or rented. The rationale is that, if left to their own devices, fishers would strip the ocean of fish. Or put another way, this system tries to stop competition among fishers because basically you can only catch a finite amount. Japanese and New Zealand governments also agreed to limit catches. In 1989, the Commission for the Conservation of Southern Bluefin Tuna was formed with Australia, New Zealand, and Japan, and since then South Korea, Indonesia, and Taiwan have joined. It decides each year on the global southern bluefin tuna quota and then allocates the quota among the participating countries.[4]

This is where Dinko enters the story again. With the introduction of quotas in the late 1980s and an interest rate of up to 17 percent, the tuna business was in crisis. Globally, ITQs always reduce the number of boats, and this happened in Australia. The number of tuna operations went from 250 to 22 seemingly overnight, as those in debt sold their quotas to others. At one point the banks owned 80 percent of the tuna businesses. The introduction of quotas was also to have a profound impact on the very idea of fishing. If you have a finite quota set in weight, not numbers, of bluefin, you do much better if you catch juvenile tuna and feed them up for sale later on. While many now say that fishers are really farmers, the temporality of the fishing industry has changed as well. Often owners catch their quota in a couple of weeks and then spend the rest of the year fattening fish and talking deals with Japanese buyers.

Dinko says he was the first to come up with the idea of cultivating tuna. Brian Jeffries, the chief executive of the Southern Bluefin Tuna Association, certainly thinks so, praising Dinko's imagination and persistence. According to Jeffries, by 1993 Dinko had perfected the technique, which hasn't changed over the years. Now tuna ranching is practiced the world over, especially in Croatia, the former home country of the Port Lincoln fishers. But the way the industry works has changed immensely since the first fishers left Kali. Fishing is now about finding fish, heavily mediated by technology. With the help of the spotter planes, the boats focus on the schools of tuna—often given away by the sea birds feeding on some of the prey the tuna have homed in on—and then cast out the huge nets. It's a bit like cattle being wrangled. The tuna are caught by purse seine nets far out to sea and very carefully transferred into sea cages. Then the cages are slowly dragged back to Boston Bay. Given they are traveling at about one knot an hour, this can take a couple of weeks. Back then, everyone thought that Dinko was mad. But in fact the idea made absolute financial, if not ecological, sense: Catching smaller tuna and then feeding them up "effectively doubled the fishermen's quota as the limit on their catch was determined by the weight of the fish caught in the wild, with any weight gained in captivity a bonus" (Lukin Fisheries 2010).

Talking to Dinko, we travel through his stories of glory and his numerous firsts. Underlying his tales of conquest is another story—one of migration, of families torn apart in divorce and greed. I sense sadness at the end of his life. His son, Dean, apparently no longer talks to him, and the sign Dinko and Sons has had the last bit painted through to read Dinko Tuna. As I leave, he asks me if I like tuna. "Yes," I say, a little hesitantly, because I'm not quite sure what he has in mind. He directs me to the back office, which is a junkyard of old computers, equipment, and boxes, and he tells me to climb up and get a box of what is labeled "John West tuna in brine." When I pass it to him, he gives it back, saying proudly that the box contains unmarked tins that are southern bluefin tuna in oil. We ate expensive endangered tuna mayonnaise sandwiches for weeks.

Watery Commons

Released from the magnetism of an aging tuna baron, I can ponder more closely the backdrop of jurisdiction and fisheries management that frames the tuna industry now. A meeting with a senior fisheries expert puts the boasting into perspective. He reminds me of what he calls "the first principles," the first of which is that the ocean and all within are common resources. This, of course, recalls Garrett Hardin's infamous essay, "The Tragedy of the Commons." In 1968, Hardin, a genetic biologist, debated the consequences of population growth in terms that are instructive and provocative: "It is fair to say that most people who anguish over the population problem are trying to find a way to avoid the evils of over-population without relinquishing any of the privileges they now enjoy. They think that farming the seas or developing new strains of wheat will solve the problem—technically" (1968, 1243).

In other words, we want our cake and to eat it too; we want to ensure a future for tuna and to eat them too. Hardin's response wends through a critique of the United Nations' Universal Declaration of Human Rights, which a year before had framed the family—its size being most crucial—as "the natural and fundamental unit of society," to which Hardin declares that the "Freedom to Breed is Intolerable" (1968, 1246). He finishes with the argument that "social arrangements that produce responsibility are arrangements that create coercion, of some sort." Hardin's pithy phrase is "Mutual Coercion, Mutually Agreed Upon" (1247).

This is precisely the structural framing of fisheries management, which began in 1968. Protecting fish comes down to enforcing measures of input versus output. It's all about controlling the number of dead fish. Fisheries management attempts to ensure that extraction will be balanced by regeneration. Input control seeks to stem the diminishing stock by controlling the size of nets and delimitation of fishing grounds and seasons, whereas output control is legislated through the quota system, described above. As a tool of conservation, it is hardly perfect: As Gilsi Pálsson and Agnar Helgason (1995) argue, it can also be a mechanism that privatizes the seas. It works in a fitful way. From 2012 to 2015, the quota steadily increased. This was naturally received with jubilation in Port Lincoln, but the biomass stands somewhere between 5 and 9 percent of what it was in the 1960s. It's hard to think that this is sustainable—for the fish and for the fishing communities.

With human-tuna-fisher sustainability in mind, I wonder about the younger generation. According to many this group includes the best skippers, with a depth of enskilment and tacit knowledge passed on from father to son—but never to daughters: Women are believed to be unlucky on board. I meet Rick Kolega and Semi Skojarev at the bar in the fancy new marina outside of town. Sam Sarin, who is the number one owner of tuna quota, built it. Number two is Šantić, and number three is Hagen Stehr, who spends his money on scientific experiments hoping to ensure a man-made future for the fish. In terms of the amounts: Sarin owns a bit over 2,000 tons, or 40 percent of the Australia quota; Šantić has 1,200 tons, and Stehr 700. It's hard to give an exact account of how much that is worth as the price fluctuates so much each year, but let's say it's between $110,000 and $180,000 per ton. Rick and Semi's company SEKOL has 10 percent, so they're not starving either. As the local paper reported, "The most outspoken of the tuna kings, German-born Hagen Stehr, likes to boast that when he sits down for his Thursday get-together at a Port Lincoln cafe with his old mates Sam Sarin, Mario Valcic and Joe Puglisi, they are worth more than a billion dollars between them" (Debelle 2006).

Sitting in the marina bar, it's clear that no one is starving. The place is flush with yachtsmen and their big boats, trophy wives, and buckets of money. Rick and Semi, who also came from Kali, are keen to recount their families' stories. Semi's father is Dinko's first cousin. Sitting with a glass of wine facing the mansions built around the marina, they're in a mood to relive the past. Rick's parents fled Soviet Croatia/Yugoslavia by boat to Italy, where they were interned. The family, including the mother pregnant with Rick, escaped by foot to Paris. From there they managed to get on a boat to Sydney, and after staying in migrant camps, they finally made the trek to Port Lincoln. Semi's father got caught trying to get out of Kali, but tried again and again until he too arrived in Sydney to eventually find his way to the tuna. Both men have been on the boats since they were teenagers, but in the late 1980s they took advantage of the introduction of quotas to buy and merge with others to form their company, which like all the others is completely focused on the Japanese market. With Dinko and other Croatian-Australians, they went back to post-Soviet Croatia to set up tuna farming there. Some see this technological transfer as unpatriotic. Others comment on the volatile mix of ethnicities, fish money, and egos.

Arguably the most famous face in tuna is that of Hagen Stehr, the owner of Clean Seas Aquaculture, which is now experimenting with breeding tuna on land. Stehr is a larger-than-life type of guy and rumors abound: that he slipped off a ship as a member of the Foreign Legion, fell in love with a local girl, and made his first million in ways that don't get talked about—much. Since then he has poured millions into a joint venture with the Japanese and with Kindai University. His idea is to breed tuna larvae into fingerlings at the Arno Bay site, which then can be raised into fully grown fish. His eventual plan is to do this all on land. He has had more failures than successes, and for many his idea is in the realm of science fiction. However, the venture is producing bred tuna that is fattened in ocean pens. In conjunction with Kindai University and A-Marine Kindai, they have brought fully artificial tuna onto the market. The fish comes "with an 'authenticity declaration' covering its life history, from its 20 days of gestation to its diet in an open-water facility to the material on netting surrounding it" (Barron 2008). They are of course a small operation compared to the big tuna farmers and brokers. The Japanese car company Mitsubishi is the world's biggest bluefin tuna broker. They have a large amount of frozen tuna in stock to be released as the price goes ever up and the remaining bluefin stock plunges ever downward. When the last bluefin is caught, Mitsubishi will have stocks of the extinct fish.

The End of Fast Fish

As the debate and politics around the tuna industry become ever more heated, food businesses in some parts of the world seem to be turning away from bluefin tuna. With the growing prominence of documentaries such as *The End of the Line*, food chains now proudly announce that their tuna sandwiches are sourced from "responsibly fished" skipjack. In upscale enterprises such as Nobu, you can still order bluefin sushi. And it is readily available in Japan. There is no doubt that the Japanese are regularly blamed for causing the demise of tuna stocks. Japan is, as we've seen, central to the value of bluefin tuna—they are what Issenberg calls "the brokers of taste" (2007, 247). But Bestor argues that bluefin tuna are more than just a commercial commodity in Japan. They are integrally linked to "cuisine, work, gender and class identity" (2004, 37). They have "cultural biographies" (131).

Would that we could have our tuna and eat it too. The solutions that are regularly trotted out tend to be simplistic and often racist. Poor West African nations are often caught in a double bind. Emerging still from colonial rule, which favored the West in everything including fishing quotas, they argue that they shouldn't have their quotas reduced. They can sell their quotas to developed nations for badly needed currency, but the amount of illegal fishing means that their seas are being fished out. In addition, the subsistence fishers are finding less and less inshore catch as the megatrawlers scoop up forage fish to feed the tuna in pens around the world. At the 2010 meeting of CITES, the Convention on International Trade in Endangered Species in Qatar, fury (mainly rather self-righteous anger from developed Western nations) was directed at Libya and Japan. Japan's lobbying may have foreclosed a motion for a moratorium on the fishing of northern bluefin tuna. However, as we have seen, bluefin tuna constantly travel the world, and it seems hard to believe that one single country can be solely to blame in this global trade. While most tuna goes through Japan, much of it then goes on to sushi bars around the world. As Bestor writes, tuna from North America or Australia is "quickly exported to Japanese markets, and is sold to brokers at Tsukiji who specialize in top-of-the-line exports, who immediately ship it back to North America, in an elaborate circulation of the cultural capital of Japanese markets, which validates the economic value of North American consumption" (2001, 83). In other words, it is at least a two-way street. Japan has taught the rest of us to savor the taste of raw tuna flesh. In a twist on the *Economist*'s Big Mac index, it could be said that sushi bars have now become a sign of prosperity and an index of the level of gentrification.

I Swam with the Tuna

And would I do it again? The answer is a resounding no. To swim with tuna is to swim in an aquarium where the species are meant for the table. Researching this book, I have visited a number of aquariums, and they are like silent zoos but every bit as desperate. The animals look so bored and so out of their element. Extend this to a magnificent beast like a bluefin, built to roam the high seas, reduced to swimming around and around in a pen. Would I eat bluefin tuna again? Apart from my sandwiches made with Dinko's canned bluefin, I've only once had the chance to eat it. During one visit to Port Lincoln, the market for tuna had badly fallen in Japan. It was the one time that you could actually buy bluefin where it was fished. Tony's Tuna had freezers full of hunks of bluefin. We bought a $10 hunk and ate it with wasabi looking out to Boston Bay.

I didn't stop eating bluefin because of the expensive, award-winning campaigns that feature fish with masks of animals we are supposed to care for and about. I wasn't put off the idea of eating tuna by the sight of Greta Scacchi in the buff fondling a tuna. If anything, the self-congratulatory and exclusive tone of these campaigns made me want to run down to the local corner store and buy some. But of course you can't buy bluefin tuna readily. Even or especially in the town that lives on it, you can only get bluefin if the export market has collapsed—which if it continued would be the end of the line for Port Lincoln. The tuna now banned in Pret a Manger cafés caused excellent publicity for the chain, but it wasn't bluefin, nor is it bluefin in the take-away sushi boxes around the world. Nobu continues to sell bluefin but has an option of albacore for the "ethically minded." The only Nobu in Australia is in a Perth casino where you can have toro with caviar for $68. My point is that the vast majority of bluefin ends up in high-end restaurants priced well beyond the means of the vast majority of people. But the reason I won't eat bluefin anymore is not the price. It is because I did swim with them. Those fish swimming around and around all day and night haunt me.

And what of the men whose stories intertwine with tuna in complex ways—the love stories of Croatian fishers now rendered farmers of the sea? When I was in Port Lincoln in 2012 for a seafood industry conference, I

Fig. 3.5. "I Swam with the Tuna." Photograph by author.

went to Dinko's office. Dark-suited men inquired of my business. I was told that Dinko had died. His sons and his wives are embroiled in legal battles over his estate estimated at over eighteen million dollars. His grandson Dean Lukin Jr. (son of the weight lifter) is leading the charge against Lukina. It seems like a sad way for the old tuna baron to go. At the conference, I sat next to Hagen Stehr. After a loss of several million, his plans to have a fully fledged commercial land-based closed-cycle breeding program are back on the go. Dinko and Hagen—two old guys who spent their entire lives living with tuna. One wanted to turn fishing into farming, to domesticate the fish; the other wants to play God-scientist by trying to bend the nature of bluefin to his specifications. Together the two tuna barons epitomize this strange industry. Driven by different desires—money certainly but also a love of tuna—these men have lived through the short history of tuna's journey from being wild to being a domesticated property of the few, to being eaten by an elite few. It's an era that will be no more, once they've

Fig. 3.6. Mick and Mario at the Ulladulla Fishermen's Co-op Society, Ulladulla, New South Wales. Photograph by author.

fished the last one. This is a tale of big masculine egos and fast fish. But it isn't only that. It's about technologies—fishing gear and freezers, planes and tuna coffins, and of the seeming compulsion of some men to go farther and farther chasing big fish. It is also a tale of men from far shores who came homing after tuna in Australia, only to lose them.

In Ulladulla I met with members of the Puglisi clan, who followed tuna from Sicily to Australia (figure 3.6). Mick's parents arrived in the 1920s, the first of what would be a long line of migrations. Mick has been central in getting the industry to fish smarter and better. His cousin Mario arrived in the 1950s and worked on Mick's boats until he could buy his own in the 1970s. I talked to Mario and his wife, Patsy, and they recounted the racism and outrage they faced when they married. They were the first interethnic couple on the coast. Patsy was the daughter of a farmer who didn't think much of fishers or Italians, and thought even less of Italian fishers. As we talked of fish and their long lives together, Mario suddenly turned to me

SWIMMING WITH TUNA   99

and said, "I have always loved the fish. Tuna is the fish I have loved." His wife agreed.

Apart from Patsy, women are silent or absent in this chapter. Locked out of the boats through superstition, caught in high-stake divorces, silently worn on the arms of the wealthy, they fade into the background of what is clearly a man's game. In chapter 4 I dive into more of the complex gender relations of fishing. Swimming with tuna certainly queered my relation to farmed tuna, rendering me queasily athwart a number of questions, one of which is gender. I hope that following the stories of fishwives and herring quines in chapter 4 will help home in on the often-silent issue of gender in the more-than-human.

4

Mermaids, Fishwives, and Herring Quines
GENDERING THE MORE-THAN-HUMAN

I send an e-mail to a colleague working in geomarine sciences about whether he'd be interested in joining an interdisciplinary dialogue about human and fish communities. His response: "I suppose you are interested in mermaids." He must have registered my e-mail signature. Of course a professor of gender and cultural studies would want to talk to a marine scientist about mermaids.

At the time his reply hurt. It felt like my passion for fish-and-human ways of being was demeaned, reduced to a little girl's whimsy. Of course my skin was already thin. I am a female professor of gender studies in an old university that, like most, values the sciences over the arts, and the traditional humanities over the come-lately areas of gender and cultural studies. But really, I told myself, being teased by a fellow academic is surely

Fig. 4.1. Mermaid in Broome, Western Australia. Photograph by author.

not as daunting as facing up to burly fishers and fishing managers. Grow a tougher skin, scale up, and fake a tail.

My colleague's comment did make me think about mermaids, for which I thank him. They are after all the perfect substantiation of more-than-human fish-women. In Western culture, mermaids have, for instance, floated through Homer's *The Odyssey*, where they appear as "singing creatures that lured enchanted sailors to death" (Romano et al. 2006, 252). The Greek sirens were women-bird-like creatures, and it wasn't until the second century BC that the first figurations emerged of them as fish-women.

Mermaids blur the real and fantastical that lurk at the edges of our psyches and in the liminal space between water and land. The mermaid is the perfect troubling figure of the impossibility of getting over gender in the more-than-human. She is slippery, hard to handle, won't or can't reproduce. She is the terrifying figure that James Cook desires when he mistakes a manatee for a woman. According to Rebecca Stott, writing about the nineteenth-century British mania for aquariums, the Victorian public was fascinated by marine zoology: "Numerous books about the shoreline and the shoreline inhabitants published between 1850 and 1880 expressed anxieties about the fragility of biological and anatomical boundaries in the animal world and the nature of animal-human kinship" (2000, 305). Stott mentions that the "marine zoological monsters of the *fin de siècle* are almost exclusively male" (306).

Of course the most famous Victorian mermaid story is Hans Christian Andersen's "The Little Mermaid," first published in 1837 and since written into little girls' lives through Disney. The original is a tale of sadness and longing. The mermaid gives up her tail to follow a prince, though the price for her new legs is to be rendered mute, losing her beautiful voice. She dances for her prince on stumps that feel like she is being pierced by swords. The trade of her tail and her voice for legs was done in the hope that the prince would fall in love with her. A heavy price to pay, but then women so often make trade-offs beyond price for love, or rather the ideal of love. In the case of the mermaid, human love was promised to make her human. His love would give her a soul. Of course he marries another, a "real" woman. The mermaid dissolves into sea foam, but a last-minute deal done by her sisters with the Sea Witch allows her to roam as a "daughter of the air." As in Homer's tale, the mermaid becomes a siren: a bird-woman. The ending places the magical though tragic tale within a frame of morality: "The daughters of the air [do not have] an everlasting soul, but by their own good deeds they may create one of themselves" (Andersen 2006).

The little mermaid is bound to a three-hundred-year sentence of doing good. The ending ensures a thoroughly female fate of serving others, of doing good and being good. However, against Andersen—let alone Disney—there are many accounts of the ambivalent powers of mermaids. Lisa Mighetto recounts that on the one hand "mermaids possessed the power to calm the ocean and could rescue seafarers from treacherous waters." On the other they "were capable of vanity, jealousy, and spite and had the ability to create enormous waves" (Mighetto 2005, 532). "Mermaids could help or hinder. . . . They represented the capriciousness of nature" (533). One of the more intriguing accounts of mermaids comes from Karl Banse, a leading marine biologist, who in 1990 published a remarkable piece in the journal *Limnology and Oceanography*, "Mermaids—Their Biology, Culture, and Demise." Apparently it's not a spoof. Banse considers at some length the plight of mermaids. He describes their comportment with men: "Regarding mermaid behavior, a recurrent theme is the habit of the females to haul out on beaches (usually in pairs) allegedly to lure, then seduce sailors; their voices were repeatedly recorded as being 'irresistible.' Perhaps they lured—but the stark fact was that they then drowned the men and devoured their flesh. Similarly, when ships broke up in gales, the females pulled sailors down into their abodes for further disposition" (1990, 150).

In this chapter, I start by asking what the figure of the mermaid can afford against the overwhelmingly masculine framing of the relation of fish and human, where the latter is always male. What does she do? To continue with Annemarie Mol's (2013) directions for researching "relational materialities"—not questions of agency—what type of "ontonorms" does she upset?

As Jesse Ransley writes in her aptly titled article "Boats Are for Boys," "Seafaring, both historic and prehistoric, fishing, trading, exploring and colonizing—the business of boats, ships and the sea . . . are 'manly struggles'" (2005, 621). Even a cursory examination of contemporary advertising representations of fish leaves one under no illusion that fishing is about men—a "manly struggle" at sea waged against fish. Whether it be the young perfectly cut fisherman competing with a black bear for a salmon destined for John West ("John West, only the best"), the generations of folksy sea captains selling Birds Eye fish fingers, or Heston Blumenthal heroically facing Arctic conditions to get Waitrose a putatively sustainable cod, the face of fishing is white and male. There is also the seemingly casual sexism of supposedly progressive organizations like the World Wildlife Foundation. As I've discussed earlier, one of its widely used slogans is "The future is man-made."[1] It seems baffling that campaigns for sustainability routinely ignore the gendered dimensions of this complex issue, baffling because it is women—who are still the main procurers and preparers of food—to whom advertising messages should be addressed. Framing the fisheries in totally gender-blind ways that naturalize fishing as an entirely masculine enterprise is part of the problem, not the solution.

It's not just advertisers and green NGOs that have a problem with gender. Within some academic circles concerned with the environment, there seems to be a pervasive sense that compared to ecological catastrophe, gender is old hat. For feminists, this of course comes as no surprise. It is part of the routine ways in which seemingly every major social movement has constructed so-called women's issues as less pressing: "Yes, dear, after the revolution we will get around to gender equality." Yet at every turn, women are differentially affected by the man-made disaster of the Anthropocene, just as they are excluded in the righteous framing of the men who will save the environment. Timothy Morton's "hyperobjects" might be a case in point. His concept seemingly encompasses everything—"things that are massively distributed in time and space relative to humans" (Morton 2013, 1). In Ursula Heise's pithy review, "It doesn't rule much in or out by way of thinking about how we live in and with more-than-human environments" (2014). Implicit to Heise's comment is that the über threat does not cast its shadow equitably upon us all.

Aaron McCright, an American social scientist, crunched the numbers of seven years of Gallup poll data and found that "women express more concern about climate change than men do. A greater percentage of women than men worry about global warming a great deal (35% to 29%), believe global warming will threaten their way of life during their lifetime (37% to 28%), and believe the seriousness of global warming is underestimated in the news (35% to 28%)" (2010, 78). Of course these data are extremely thin about how women differ from each other in terms of how race, ethnicity, class, and geopolitics intertwine with gender. Yet again this reflects the fact that those differences don't count enough to be counted.

Even when the gendered dimensions of climate change are acknowledged, the framing is extremely limited and limiting. As Seema Arora-Jonsson states, "Dual themes recur throughout the existing though limited literature on gender and climate change—women as vulnerable or women as virtuous in relation to the environment" (2011, 744). This binary obviously mirrors others: Women in the Global South as victims of poverty and so-called natural catastrophes, and those in the North as virtuous eco-consumers. Arora-Jonsson is particularly concerned with how this binary is reproduced in government and nongovernmental policy: "A focus on women and their vulnerability can deflect attention from power relations and inequalities reproduced in institutions at all levels and in discourses on climate change" (2011, 745).

Gender myopia frames the space where man-made environmental change can be addressed, as well as how it is addressed. The 2014 report in the influential *Annual Review of Environment and Resources*, "Gender and Sustainability," states the case clearly: "Most programs to promote sustainability have been gender blind and thus ended up working primarily with men, who are more likely to occupy public spaces (including community organizations and government or external programs) and are often more readily recognized by outsiders as the foresters, irrigators, fishers, and even farmers" (Meinzen-Dick, Kovarik, and Quisumbing 2014, 47). In other words, through their programs NGOs and government bodies engage with men and thus re-create a public understanding that the problems are to be resolved by male players. The past, present, and future are man-made.

However, even in this report where considerable effort has gone into thinking about gender and environmental sustainability, how gender itself is framed is problematic. The authors are from the International Food Policy Research Institute, a highly respected organization that is part of the massive CGIAR Consortium—a sprawling network of international state-based organizations and global ones such as the World Bank that conduct research and run aid programs centered on eradicating hunger. Much of their work is focused on scientific research, from which they frame policy recommendations. The review opens by discussing its theoretical framing of gender, which has a heavy reliance on ecofeminism. The legacy of ecofeminism is important, but it is also only one way of viewing gender. The authors of the report define ecofeminism as framing women as inherently "closer to nature . . . biologically, socially, materially, and ideologically" (Meinzen-Dick, Kovarik, and Quisumbing 2014, 33). They argue that, from an ecofeminist perspective, "women are more likely than men to be focused on sustainability and conservation because of an inherent drive and a desire to conserve" (33). In their conclusion, the authors return to critique the "ecofeminist myth" that considers "only one side of the evidence on the extent of—and reasons for—gender differences" (48).

This is all a bit too easy. Here the authors take ecofeminism as the privileged term of feminist engagement with the environment. Having thus privileged one theoretical perspective, they then critique it as if it were the only feminist intervention. They do this with no regard to internal feminist critiques. Numerous contemporary feminist arguments have pointed to the purported essentialism of ecofeminism. As Andrea Nightingale notes, within ecofeminist discourse, "'women' was a largely undifferentiated category and it was assumed that all women would have the same kind of sympathies and understandings of environmental change as a consequence of their close connection to nature" (2006, 167). Obviously, reducing women to nature is problematic, but equally there are several strands within ecofeminism that have been and continue to be politically and theoretically important. For instance, despite her criticisms Nightingale acknowledges that Vadana Shiva's groundbreaking 1988 ecofeminist book *Staying Alive: Women, Ecology, and Development* was crucial to making the multiple connections between gender and the environment visible.

This acknowledgment is important because it recognizes intellectual debts (something that often goes missing or is rendered invisible by academic amnesia). It also historically contextualizes how, where, and for what reasons questions of gender and the environment became important. And it gestures to the uneven ways in which ideas are taken up and circulate. We all have historical intellectual trajectories, which limit or enable our access to certain forms of ideas and knowledge. Sometimes these are idiosyncratic. For my own part, I did my undergraduate degree at the University of British Columbia in Vancouver in the mid-1970s when ecofeminism and affiliated discourses reigned. I ran in the opposite direction—to Foucault and so-called high theory. In my youthful and probably addled mind, ecofeminism was another strand of hippy-dom. As a queer brat, lipstick and disco were my forms of attachment—not folk music and earth mothers. I was more into rockin' in the cosmos than being at one with the world.

I now realize how narrow my understanding was. To return to Nightingale's nuanced understanding of the issues and the times, ecofeminism figured a symbiotic bond between women as biological beings, and the earth as biological and ecological systems. This is not the same as equating women with nature. Nonetheless this attracted charges of essentialism both then and now—as in the example of the NGO review. One has to remember that "essentialism" was the favored insult directed by various women's groups at others, a form of feminist shaming (Probyn 2005a). Some of that shaming was deserved and politically generative. Critiques by black feminists made visible how some white women essentialized the women's movement through their projection of their own particular issues onto the larger movement. Middle-class white women (who were often in charge of the show) also ignored the differential experiences of working-class white women. Needless to say, even in the recent statistics to which McCright draws attention, no differentiation is made based on women's class or ethnic belonging and how that might or might not affect how they experience environmental change. For instance, most urban farmers' markets speak loudly of the continuing middle-class bias, and are a prime site to see how environmental virtue is performed and reproduced. Yet as far back as the early 1990s, feminists such as Bina Agarawal argued that class was a crucial factor in understanding women's relationship to the environment.

While at times ecofeminism may have privileged a relation between (undifferentiated) women and (equally undifferentiated) nature, I am more concerned here about how institutional organizations take up certain ideas about what gender is. For instance, the "Gender and Sustainability" report concludes that "gender is closely linked to biological sex" (Meinzen-Dick, Kovarik, and Quisumbing 2014, 35).

Well, not that closely. Some twenty-five years after Judith Butler published her groundbreaking *Gender Trouble*, which rewrote how we understand the relation of gender to sex, it seems extraordinary to hear this claim. Butler's profound charge was that "gender is not to culture as sex is to nature; gender is also the discursive/cultural means by which 'sexed nature' or 'a natural sex' is produced and established as 'prediscursive,' prior to culture, a politically neutral surface on which culture acts" (1990, 7). This oft-cited argument remains important, and perhaps ever more so, because the conflation of gender and sex is what allows for the formulation of women and the environment as "natural," for example, as not important politically. Gender is not a thing; it is a relation or more accurately a presentation of a relation. In this I am indebted to the feminist film theorist Teresa de Lauretis and her formulation of gender as "the representation of a relation, that of belonging to a class, a group, a category" (1987, 4). She continues, "Gender represents not an individual but a relation, and a social relation: it represents an individual for a class" (5).

In this way, gender is not women or men but rather the social operations that constitute us in a binary of M:F. This is not to posit a universal sexual difference, but rather directs us to how we come to be represented as belonging to the class of woman or man. That process (which, following Louis Althusser, de Lauretis calls interpellation or "hailing") is always culturally and historically specific. "Gender" is a dynamic term that is constantly in play in the dialectics of producing us as variously feminine or masculine or trans, and most importantly, these categories are given meaning and concrete life by how they are inhabited and embodied. We are not statically placed within a monolithic category. We are always subjectively refiguring and reshifting ourselves as gendered in one way or another. As de Lauretis puts it, "Gender both as representation and self-representation is the product of various social technologies, such as cinema, and of institutionalized discourses, and critical practices, as well as practices of daily life" (1987, 2). Every time an official report such as the "Gender and Sustainability" review states that women are better at sustainability because of their closeness to nature, we are in the presence of a certain ideology.

Enter the More-Than-Human, Exit Gender?

Across my academic life I have experienced a lot of "posts." The arguments of Butler, de Lauretis, and others are often tagged as post-structuralist. This in turn overlapped with the moment of postmodern theory. Even when I was actively engaged in the thick of often-acrimonious debates for or against such theories, the lack of specificity of these umbrella terms annoyed me. In turn, queer theory took on many of the same authors who had been categorized as post-structural or postmodern and turned them to undoing "gay and lesbian." Most notably Foucault figured in this, as did Butler. And then of course there was the magnificent Eve K. Sedgwick, whose understanding of queer as athwart runs through this book.

Alongside these twists and posts, a great amount of energy has been directed at destabilizing the sanctity and the separation of the human from its ecological context. In her early review of the connections and disconnections between feminism and human-animal studies, Lynda Birke argues that "discontinuity is reinforced implicitly, and the chasm yawns between human culture and the rest of nature" (2002, 430). She cites Mette Bryld and Nina Lykke's important point: "'Human' is definitely not a neutral or innocent category, but a highly gendered and racialized one" (Birke 2002, 430). Taking up the legacy of groundbreaking African American feminists such as Hortense Spillers, Alexander Weheliye pushes this to a new threshold, arguing that "analyses of racialization have the potential to disarticulate the human from Man, thus metamorphosing humanity into a relational object of knowledge" (2014, 32).

Gender continues to be important in feminist science studies, an area that has greatly contributed to studies of the post- or nonhuman. But what gender does becomes very diffuse. Within the camps of the posthuman and the nonhuman, the objects of study are exotic. For instance, in Karen Barad's (2012) work, her "queer critters" are atoms, dinoflagellates, and stingrays. Barad (2012, 29) works to profoundly destabilize boundaries by interrogating the division between the human and the nonhuman. There is a busyness along the crevasses of these boundaries with Eva Hayward's (2012) "sensational jellyfish," the various strange critters within the multispecies encampments, and Jane Bennett's (2009) "stuff" of "vibrant matter." But sometimes my head begins to buzz with the ferocious pace of word play and theoretical riffs. There is also something disturbingly blithe about

Morton's gleeful account of his hyperobjects, even as it breezes across one of the most disastrous oil spills in history: "The BP oil spill of 2010 provides yet more evidence that ecological reality contains hyperobjects: objects massively distributed in time and space that make us redefine what an object is" (2010, 167). In this theoretical busyness, I lose sight of the crucial baseline of gender and queer: of the kinds of embodied engagement, the lived relatedness of stuff that matters.

However, these terms—the "more-than-human," the "nonhuman," the "posthuman"—are generative in that they seek to shake up any assumptions that we might have had about what conjoins and what separates us, not to mention what that profoundly confusing "us" might be. As Whatmore puts it, the onus needs to be on "startling our habitual assumptions about what life is" (2004, 1362). Like Whatmore, I prefer the term "more-than-human" to "posthuman" or "nonhuman."[2] It is, as I argued about the term "gender," ontologically and materially relational, and opens up new epistemologies as it narrows the diverse and shifting relations between and among humans, and the many different aspects of that are so much more-than-human. For feminist geographers such as J. K. Gibson-Graham, it is, or it could be, "a vital pluriverse, suggesting an affect of uncertain excitement, an ethic of attuning ourselves more closely to the powers, capacities, and dynamism of the more-than-human" (2011, 3). It is a space that should caution modesty, which could call forward "an ethics of attunement and a more sensitive, experiential mode of assembly" (3).

As Donna Haraway writes in *When Species Meet*, contact between and in the more-than-human is as fleeting as a touch. But it is also as large as what touch might inaugurate: "Touch ramifies and shapes accountability. . . . Touch does not make small; it peppers its partners with attachment sites for world making. Touch, regard, looking back, becoming with—all these make us responsible in unpredictable ways for which worlds take shape" (Haraway 2008, 36).[3] To this we can add Nightingale's framing of how "gender is not constant and predetermined materially or symbolically but rather becomes salient in environmental issues through work, discourses of gender, and the performance of subjectivities. Not only are inequalities between men and women a consequence of environmental issues, gender is a cause of environmental change in the sense that gender is inextricably linked to how environments are produced" (2006, 166).

Here the salient point is "how gender and environment are mutually constituted" (Nightingale 2006, 170). The expansive space of that mutuality is profoundly important. As in Haraway's call to focus on connection and accountability, this is a generative framing that allows us to imagine and bring into being all sorts of co-constitutions. This entails different forms of responsibility to so many entities: fish, environment, ideas, people who become partners in research, memories. These forms of entanglement do indeed produce worlds that we don't yet, will never, fully know.

Fine Figures of Fish-Women

I like the idea of homing. Humans have long tried to understand how fish, especially salmon, home—that amazing capacity to return to the stream in which they spawned. But I'm also intrigued about how humans home in on fish. Each year over the millennia humans have turned to certain places to find fish. There is, at times, a becoming-with, a way of being that subtly echoes the fish they desire. In Bear and Eden's work on "thinking like a fish," their fishermen attempt to "become-fish that they cannot see" (2011, 350). In this section I want to distinguish how women become fishy. In 2010 I followed the ghostly trail of the long-gone fish up to the northeastern coast of Scotland. I wanted to find out what happens to communities that lose the fish, but it was also a trip of remembrance—two quite different forms of responsibility that coincided. I spent several summers when I was a girl in Findochty, just along from Lossiemouth on the Moray coast (figure 4.2). My parents would drive us there in a small car from mid-Wales, which was a long trek with two small bored girls in the back. My father and mother loved fish and fishing, and the mighty fine single malt produced down the road in the Speyside. By the time I could home my way back to the places of my girlhood summers with Scottish fish, my father was too frail to make the trip. So I said I would do it for him. He has since died, and the ghosts multiply.

At a pub in Aberdeen, I was lucky to fall in with women who had grown up in fishing families of the northeast that went back generations. Falling in with women is an interesting process. Normally it's hardly onerous. But you need to do your homework, listening, looking carefully and casually. We'd all escaped the confines of the stuffy pub. It was raining, of course, and hard, so we gathered under the little atrium and talked while some of the women smoked. When they heard of my interest in fishing and women, they beckoned me to their homes by the depleted sea. So I followed these fish-women several hours up the road to Lossiemouth. They spoke easily of

Fig. 4.2. Findochty, Scotland, in fog. Photograph by author.

love for the fishing life and deeply of the seemingly concomitant loss and sadness. "Death is omnipresent," said Sheila, thinking of three lost sons. Sheila, the great-great-granddaughter of Lossiemouth fishers, was especially welcoming, and we sat in the sun house at the back of her home drinking wine—and smoking. She was close to retirement, and after we had talked, she said in a quiet and determined way that she'd like to write down her own stories of fish and the life that was "before greed made men hoover everything out of the sea beds." I hope she has.

Before I follow Sheila to find the historical and physical spaces where fish and women commingle, let me draw breath and look back on the story so far. We have encountered the figure of the mermaid, the fish-woman par excellence, trapped in rough-hewn sets of meaning. But perhaps she isn't trapped after all; maybe she can jolt us out of routine ways of thinking about human-fish being. According to Rictor Norton, Alfred Kinsey, the American sexologist, was fascinated by Andersen's obsession with the mermaid.

"Kinsey could say unequivocally that they were straight-out homosexual stories; like the mute Little Mermaid, Andersen could not tell the world of his own homosexual love for the people of the world, but the original manuscripts showed his feelings clearly" (Norton 1998, 129). Here the more-than-human fish-woman offers Andersen succor, and offers her body and her voice as a cipher for an impossible love. She is, after all, a woman incapable of reproductive sex, her tail too easily standing in for a phallus. She is also Thalassa, the sea goddess that Shé Hawke (2013) dubs "the aquamater." She stands in rebuke to those who would diminish gender to a mere echo of the sexual. Against the cute Disney representations of the mermaid, Banse's scientific fish-woman-beings await their turn to lure and devour the flesh of human men. Fish and woman, she is the figure that prompts "laughter . . . in the realization that all along the original was derived" (Butler 1990, 139). Here the power of gender as a queering relational term promises to disrupt the somewhat flat equation of the more-than-human.

Let us see what light these contradictory elements may shed as we follow women following fish. This is a tale that of necessity combines the fantastical in and with the everyday experience of women in the rough seascapes of northeastern Scotland.

In 1860, the traveler and historian Charles Richard Weld came across young fisher girls. They obviously made an impression. He described them in his book *Two Months in the Highlands, Orcadia and Skye*: "The herring lassies covered in fish guts . . . [so] 'bespattered with blood and the entrails and scales of fish as to cause them to resemble animals of the ichthyological kingdom'" (quoted in Nadel-Klein 2003, 81–83).

Jane Nadel-Klein, whose book *Fishing for Heritage* contains much wonderful historical description of the fishing industries in Scotland, says that from Weld's description "one could even infer his belief that if Darwin's theory of evolution were correct, the primitive fishwife was working her way back down the evolutionary ladder" (2003, 83). Here she refers to Weld's critique of *On the Origin of Species*. Weld comments, "If a man may become a monkey, or has been a whale, why should not a Caithness damsel become a herring?"

In the nineteenth century, these Scottish herring-damsels broke loose from their traditional roles tending the nets and followed the herring. The women were called the herring lassies or, in the Doric of the east coast of Scotland, herring quines. They and the fish performed an amazing yearly homing. "Traveling with the fish," as Sheila put it, they were an important part in the more-than-human assemblage of fish, institutional power, and technology.

Before we follow that tale, let me set the scene. Traditionally, the heartland of the Scottish fisheries was the strip of coast from Inverness on the west to Fraserburgh and Peterhead on the east. The small fishing villages along the northeast Moray coast, such as Lossiemouth, Buckie, Banff, Findochty, and Portsoy, were called fishertouns, and fishing was all there was—along with the "Kirk." Salmon, haddock, and cod had been caught since the Middle Ages with small open-deck boats that used "small lines"— with up to a thousand hooks per line. In 1656, Oliver Cromwell remarked on "the herring fishing, which are caught thereabout and brought thither, and afterwards cured and barreled up, either for merchandise or sale to the country people who come thither, far and near at the season, which is from about the middle of August to the latter end Sept" (East Lothian Museums 1999). The women's role was to bait the hooks after they had gathered the bait—sea worms, mussels, and limpets. Many report that women's lives as fisher lassies were hard and that they were remarkably strong women. A traveler in the late nineteenth century reported that "the men were for the most part watching their women-folk at work. They were to an astonishing extent mere spectators in the arduous task of hauling the cobles [the open fishing boats] one by one onto the steep banks of shingle. . . . With a last 'heave-ho!' . . . the women laid hold of the nets, and with casual male assistance laid them out of the shingle, removed any fragments of fish, and generally prepared them for stowing in the boats again" (Frank 1976, 69).

In addition to hauling the boats, one of the main duties of the women was to bait the hooks. We are talking about a lot of bait. "As a crude guide," Peter Frank writes, "for a single night's fishing the number of hooks to be baited for a three-man coble ranged from a minimum of 2,600 . . . to a maximum of 3,360" (1976, 61). And of course the bait had to be freshly caught. Mussels would need to be prized open and then caught on the hooks. If the weather turned and the boats didn't go out, the women had to do it all again the next day. It was tough work, and the women had to wear feminine garb: "Ankle-length dresses over a quilted petticoat offered little warmth or protection from the chapping wind; but when they got wet from the rain or the tide, as they almost invariably did, the sodden, heavy material flapping against their legs must have caused intense discomfort" (Frank 1976, 63).

Because there were few proper harbors where the boats could moor, the women would carry the men through the water to their boats. Sheila told me that it was "so the men wouldn't get their feet wet." There is contention about this practice: Was it men lording it over their women, or was it a practice based on a commonsense calculation conducted by the women? At the time, fishing boots were made from hard leather and the boats had no cover, so the men's feet would have been sodden, heavy, and potentially dangerous during the long fishing day and night.

In the early 1800s, the Church, which was heavily involved in fishing, including demanding their tithes or *teinds* in fish, decided that the boats should also go out in summer months when the cod were no longer in season, but lesser fish such as herring were plentiful. According to Rosemary Sanderson (2008) of the Banffshire Maritime Heritage Association, the Church would distribute the nutritious and cheap fish to the poor. The more moneyed classes wouldn't touch herring. However, the herring market soon developed with exports to Ireland, Germany, Russia, and even the West Indian plantations, where salted Scotch Cure herring was cheap food for slaves. Herring, pilchards, and sardines all belong to the Clupeidae family. They feed on plankton and live in large shoals, making them easy prey for fishers. The herring swam—and to a limited extent they still do—from the top of the Shetlands and down along the northeast coast, and then south to Yarmouth. That is a long way: some eight hundred miles of wild seas. Because herring begin to deteriorate within twenty-four hours of being caught, they need to be quickly gutted and cured. This is where the fishing lassies came in. The girls, as young as thirteen, followed the boats that followed the "silver darlings," as the fish are still called. Stanley Bruce's poem gives a sense of what the lassies faced: "The sea win' cud be bitter / stan'ing at the farlan's / but the lassies they aye stuck the gither / fa'n they followed the 'Silver Darlings'" (Sanderson 2008, 11).

The Lives of a Quine

It was a tough life following the silver darlings, but compared to their daily drudgery during the winter baiting the lines, mending the nets, and carrying their men to fish, it did offer some excitement. The season started in May, and the girls moved slowly southward until the end of the season in late November. At the peak of the herring industry, 12,500 herring quines were working, of whom 5,000 would travel all the way down south to Falmouth. The herring catch started up north around the Shetlands. The lassies traveled on third-class rail and on ferries and lived in huts owned

Fig. 4.3. Sign for the Silver Darling Restaurant, Aberdeen, Scotland. Photograph by author.

by the curers, who were in charge of most aspects of the herring trade. The quines worked twelve hours a day or more, except on Sundays. Their skill was remarkable: The gutting crew would gut the fish with a technique called gibbing, whereby the throat, gills, and guts were removed in one movement, all with their hands protected only by strips of flour sacking known as clooties. The packing crew would clean the fish, salt them, and pack them in large barrels. A good packer could pack up to twelve hundred herrings in ten minutes, and thirty barrels a day. That's thirty thousand herrings per girl each and every day, except of course on the Sabbath (Sanderson 2008, 20).

It may have been a hard life, but from the accounts, photos, and paintings that adorn the many maritime museums along the coast (often the sole remnants of this once-global industry), the women looked happy to be away, free from home, and glad not to be in service. As Sanderson writes, they had an "on the move lifestyle" complete with "hand-fasting." This ancient Celtic practice was revived in the eighteenth century as a trial marriage, which the woman could break off if she had not become pregnant within a year and a day (Sanderson 2008, 60). In Nadel-Klein's astute reading about how fish-women negotiated the "gendered face of stereotype, stigma, and marginality" (2003, 51), it's clear that the fisher lassies and herring quines occupied an interesting position in what was a staid society. One of their most striking traits was that they were seemingly less concerned about respectability. Nadel-Klein writes that "women's prominent, public roles in the fishing industry helped to brand them and their families as odd, indeed, as less respectable than most other Scots" (2003, 51). Of course, as Valerie Burton points out, the very term "fishwife" carries "misogynistic intent" (2012, 527). "The women who dealt with fish [were seen] as brash, brazen, and unclean . . . the stench of rotting fish [standing] in for the odors of female genitalia" (531–32). Perhaps especially in northeastern Scotland, where the hand of respectability lies as heavy as the pall of the haar, the heavy sea fog, the chance to get away—to "follow the drifters"—must have been very welcome.

In Paul Thompson's comprehensive social history of the Scottish fishing industries, *Living the Fishing* (Thompson, Wailey, and Lummis 1983), the quines certainly seem to like getting away and being in the company of other women. According to Thompson, "They formed the crews [of three] by themselves at home, with friends: they put their heads together" (1985, 9). As one girl remarked, "I was free—because once you were off fishing in the year the time was your own and you'd nobody to say, well do this, you see—you were just with the girls; and you would go to the theatre, you could go to dances or anywhere" (Thompson 1985, 11). If the image of a fishing lassie "bespattered" with herring guts going to the theater in Shetland seems extraordinary, there were other goings-on that would have been even more shocking. Apparently for some, "the freedom brought revelations." Christian, one of Thompson's interviewees, remembers when she was a girl of sixteen in Shetland. A mainlander, she was amazed at the carrying on of the women from Wick. They lived six to a cubicle in shared huts. It was easy enough for the young girl Christian to climb onto the rafters to spy on her fellow fishing lassies. Probably hoping to catch them in the act with the herring fishermen, she was surprised to see "what was goin' on in the next hut—they were nae in their beds until it was rising time! We went into the next . . . and we looked over—they did nae occupy their time with men at all!"

The Wick women had obviously found other women more to their liking than the probably fairly rough and ready fishers. Christian exclaims, "The Wick women—oh! What a time they had!" She concludes, "We finished our education, eh?! Ach, it's true—what we didna ken, we learnt then!" (Thompson 1985, 12).

Thompson is a social historian (not a gender studies scholar), and he rather quickly lets this intriguing episode pass in favor of his thesis about what the fishing lassies reveal about labor relations. The Scottish fishing industry is indeed a fascinating microcosm, especially for labor historians. The industry emerged in its more formal state as a direct result of the Clearances that I touched upon in chapter 2. Those who did not immigrate to North America or to the antipodes found themselves under a decree laid down by James IV of Scotland (later to be James I of Great Britain under the union of the Scottish and English crowns). His law was that "idle persons" should be pressed into the service of the fisheries. Neil Gunn, one of the most prolific writers of recent Scottish history, puts it this way in his 1941 novel *The Silver Darlings*: "The landlord had driven them from these valleys and pastures, and burned their houses, and set them here against the seashore to live if they could and, if not, to die" (Nadel-Klein 2003, 37).

The emergence of the Scottish share system continues to interest labor historians. Nadel-Klein explains, "Each fisherman received an equal share of the profits, if any, after allocating two-thirds of the catch to the boat and gear" (2003, 38). This ensured a striking equality, especially for the time. The fisher lassies feature in this system as independent players. As Thompson puts it, "The gutting girls were simply seasonal, employed wage earners. Unlike the fishermen, they had no shares in the herring trade. They thus had less to lose than the men, and, as many of them remember immediately now, they were full of spirit" (1985, 10).

It was the women who mobilized a long series of strikes from 1911 until well after World War II against the low price of herring. Because they had no shares in the system, they could strike without imperiling their fellow fishers in the shared scheme. But they also held a definite position of power. If the catches came in with no one to gut the fish, they would quickly rot. These were often "lightning strikes," where "young girls [were] egged on by one or two married women who care[d] more for the fun of the thing" (Thompson 1985, 10). The women achieved wage increases, often in pretty rough circumstances. Nonetheless "Elsie Farquhar of Buckie remembers that they all rather enjoyed it" (10).

However they are depicted—as frivolous strikers or as central to disrupting the overwhelming power of the curers—there is no doubt that the fishing quines were integral to the herring industry. But the industry itself was on the decline. It started to falter during World War I, slightly recouped in the interwar years, staggered again in World War II, and then by the 1950s was in its death throes. The initial cause of the downturn was man-made, but not in the way we would now think of it (either from climate change or overfishing, although that was to come). Rather, it was the human disasters of the two great wars that shattered the herring industry. The biggest markets were in Germany and Russia, and with the fighting the demand flattened and the supply was curtailed. The fishing lassies would have found better-paid and certainly less cold and onerous work in the munitions factories during the war. After the war, attention turned to the more valuable whitefish or demersal fish, and with advances in technology and trawling, and purse-netting replacing herring drifting, the end was nigh. The "klondykers" from the USSR and eastern Europe arrived, and these enormous factory ships soon gobbled up the remaining fish and the work of the Scottish boats. By the 1970s, the herring had gone and the industry was closed down.

Sitting in her home in Lossiemouth, Sheila and I discussed the decline of the fisheries. As the fighter jets from the local RAF base roared overhead, I wondered at the loss. Certainly it's a loss of fish, of life, and gradually the loss of the history of the connections that bound women to the fish. Sheila doesn't omit the hard parts. She married a fisherman who was the youngest skipper in the area, taking over after his father died at sea. In turn, he lost a man overboard. He couldn't get over it and was lost to the bottle and to his wife. It was always tough—Sheila said it was being "a virtual single mother when he was away and then at the weekends he was at the pub." The worst, she said, was "not being able to plan ahead."

And now there's little to be able to plan for. Later I talk to a couple of old skippers at the Buckie Heritage Centre. There's been no fish from Buckie since the 1990s. This is the fishing town that I remember from my youth, where the market overflowed with all sorts of fish and crustaceans. They remember it too, of course, and much more painfully. What's really galling to the old fishermen is that there's lots of fish to buy. It's frozen and comes from Faroe Island boats, not governed by either U.K. or EU regulations. Eric Smith, who got out of fishing in 1984, says, "I had enough. I'm nae gonna break any laws." Another speaks up: "We was brought up with herring in baskets. Now the herring is pumped out of boats. They [the Faroese] are away for a month—they make millions from the export." They both sigh: "It was sad, ya ken. It was life."

On from Buckie is Peterhead, which was once a thriving whaling port. Then it was herring central, and now it's mainly whitefish. Going down to the port, there wasn't much to see. Admittedly it was in the afternoon, but the only signs of life were the smiling Filipinos who now constitute the majority of fishing crews on the big foreign-owned boats that land at Scottish ports (much to the fury of the few remaining local commercial fishers). Figures 4.4 and 4.5 express how much things have changed. The older white working-class members of the Buckie Heritage Centre are in stark contrast with the Filipino crew. For the Scots, fishing was a life that built community, whereas for the crew from the foreign trawler, fishing for eleven months of the year allows their families back home to survive but does little to build community there. The two Filipinos were resigned to it. As one joked, "[Labor] is the greatest export of my country."

While the Peterhead continues to be important for whitefish, the herring are gone. Also gone are most of the inshore fishing boats. Now it is just the supertrawlers that, as the old guys said, "hoover up the fish." They can process at sea and rarely need to connect with the shore—and with the people whose lives used to be entangled with the fish. Those fishers left at sea make a living on shellfish, whose numbers ironically are now quite healthy. As I explained in chapter 1, once the carnivorous fish have

Fig. 4.4. Volunteers at the Buckie Heritage Centre, Scotland. Photograph by author.

Fig. 4.5. Filipino fishing crew at Peterhead. Photograph by author.

gone, lobster and langoustines have fewer predators and their populations explode. In the short term, they're a valuable catch that is flown direct to European cities. In the long term, it's a worry on every count. Fishers have to go further out as the shellfish are over toward the west. They have had to invest in faster boats, which are less stable, and as Penny Howard (2015) writes from her fieldwork with fishers on the west coast, the mortality rates have risen steeply. It's a worry for the seabed as the more commercially viable gear is trawling, which rips up the seabeds. It's a worry for the local communities because this highly valued commodity is flown out to export markets. This means little employment is generated in small towns like Peterhead and Fraserburgh, where multigenerational unemployment is high.

Who is to blame? There is so much vituperative posturing. There's so much noise and little sense of which parts of the problem connect to others. Nightingale argues that the continuing ways in which fish are constructed solely in biological terms is at the heart of the problem: "Fisheries management remains firmly embedded in a logic that frames humans and fisheries as separate and distinct phenomena, relying upon scientific assessments of fish stocks and the mapping of sea beds to ascertain the impact of fishing activities on the biological communities of fish" (2013, 2363).

A Man-Made Crisis?

One of the most devastating moments in modern fishing happened on July 2, 1992, when the Grand Banks of Newfoundland were closed to cod fishing. They were one of the most fecund fishing grounds that recorded history has ever seen. As Dean Bavington writes, "In 1497 John Cabot reported codfish so thick they reportedly slowed the movement of his ship the *Matthew* on the Grand Banks." Three hundred years later, these were grounds that Thomas Huxley said were "inexhaustible" and "the destruction effected by the fisherman cannot sensibly increase the death-rate. . . . Nothing we do seriously affects the number of the fish" (Bavington 2008, 99).

The Atlantic fisheries connected the New World and the Old World geographically, economically, and culturally. Newfoundland played an important part in this New and Old World fishing connection. The wild and isolated island drew humans who homed in on fish. In Newfoundland, it was the Irish who crossed the northern Atlantic in large numbers: "Between 1800 and 1830, over thirty-five thousand people from a region of Ireland with a radius of about fifty miles settled in sparsely populated Newfoundland" (Dunphy 2013). These were people who had been farmers but came to love the life of fishing. When the Canadian government passed the moratorium on cod, "50,000 workers were displaced from the fishing sectors, and nearly 50% of harvesting plants were labeled redundant, and the economies of hundreds of communities were decimated" (Davis and Gerrard 2000, 280). In 1992, the total population of Newfoundland and Labrador was 580,000. Now it is closer to 510,000.

In the aftermath of the closing, media and government focused narrowly on the plight of the fishermen. I was living in Montreal at the time, and I remember vividly how the CBC (the Canadian Broadcasting Company) wallowed morosely, wondering about what the men would do. Dona Lee Davis argues that the closure caused a "feminization of local men." It's not surprising that the men were at a loss. Some left, but "many of those who [had] chosen to stay [sat] in the bars literally and figuratively crying into their beer" (Davis 1993, 469). Davis calls this "the ruined gender," arguing that "a lack of access to a maritime livelihood has challenged the very roots of masculinity. Men on land become like women. In the late 1970s a man who would not fish and spent much time in his home was referred to as a biological or psychological anomaly, 'more like a woman than a man'" (473).

Other feminists argue that the crisis occurred partly because every aspect of women's roles and lives was ignored. In Newfoundland the women were most often employed as fish processors, and they would also keep their husbands' accounts. They were in a perfect position to note the changes in the size of the fish and the gradual diminishment of the catches over time. But no one asked for their opinion. And when they spoke up, no one listened. As Nicole Gerarda Power puts it, "The marginalization of women processing workers within fisheries science and management has limited our understanding of how capitalism and male dominance have interacted to create resource degradation and shape our responses to it" (2000, 203). Barbara Neis, one of the pioneers of a feminist analysis of the fisheries, is equally blunt: "Gender relations permeate fisheries at every level." And women's ecological knowledge has been mediated "through their relationship with men—fisher*men*, husbands and sons, male-dominated governments, and male-dominated science and industry" (Neis 2005, 7).

One might have hoped that the colossal shock caused by the closing down of the cod fisheries might have triggered some self-reflection on the part of these male managers and government officials. However, according to Power it has further entrenched a blind ignorance to women's experiences and knowledge. In the aftermath, the Canadian government's policy was to professionalize the industry. In practice, this meant the introduction of strategies that "enable a downloading of the responsibility for fisheries management onto an exclusive group of self-reliant individual entrepreneurs" (Power 2005, 102). In other words, this was a clear-cut exercise in neoliberalism, and of course men were the ones both to manage the change and to be trained to be more professional. It is not for nothing that this policy was introduced under the government of Brian Mulroney, which lasted from 1984 to 1993. Mulroney could be called Canada's first neoconservative prime minister. The government actions went hand in hand with cuts to family and community support. Their further dismantling of Newfoundland's fishing communities was aided by the view that central Canada has of Newfoundlanders as dimwitted "Newfie" dole cheats.

Shutting the door after the horse had bolted, the government introduced individual transferable quotas (ITQs), which, as we've seen, often serve to privatize the commons. In Power's words, "'Ownership' of resources does not necessarily mean conservation" (2005, 103). More specifically, as we saw in the case of the tuna licenses, it translates into the concentration of quotas in a few hands; often men who do not even live in the local communities. In Power's argument, "when fishing rights are attached to ownership of fishing 'property,' the patriarchal dividend is upheld or created since men tend to be the ones who formally have ownership" (2005, 103).

The French fisheries activist Alain Le Sann describes fisheries "as one of the most highly globalized economic sectors" (Neis 2005, 2). This has always been true to an extent—in the north of Australia, for instance, Aboriginals traded in fish and pearl shell with the Makassar seafarers of Southeast Asia centuries before the Europeans arrived (Balint 2012, 545). However, with the widespread introduction of ITQs in the Global North and the admonition by the World Trade Organization that countries in the Global South should use their fishing rights in debt repayment, the stakes are increasingly high. This is particularly acute in Africa and Asia, where fish are the main source of protein and fisheries the major source of livelihood for hundreds of millions of people. The marine ecologist Daniel Pauly lambasts what he calls the "fishing-industrial complex—an alliance of corporate fishing fleets, lobbyists, parliamentary representatives, and fisheries economists." He adds that "by hiding behind the romantic image of the small-scale, independent fisherman, they secured political influence and government subsidies far in excess of what would be expected" (Pauly 2009).

Not surprisingly, what Pauly calls an "Aquacalypse" hits women particularly hard. A report by Lorena Aguilar and Itza Castañeda of IUCN-Gender and Environment draws on the UN's Food and Agriculture Organization's estimates that worldwide 35 million people are directly engaged in fishing activities to argue that in the Pacific region alone, it is estimated that women catch about a quarter of the total seafood harvested. Worldwide, women are overrepresented in fish processing (Aguilar and Castañeda 2001). Something like 90 percent of fish processors are women. The work is long and arduous and often overlooked as a source of knowledge. The numbers are really only estimates, as women are simply not accounted for across the fisheries. Meryl Williams (2010), a leader in gender and fishing research, has long argued for the need to have better gender-disaggregated statistics in order to more effectively implement development plans centered on women and fishing in the Global South. Again, it's the case that if women are not counted in ways that end up in the statistics, they disappear from sight. Then they do not have to be accounted for. Like Pauly, Williams has worked for WorldFish, an international research organization. WorldFish works primarily in Asia, Africa, and the Pacific. It is unusual in that it includes social science along with marine science and it has a primary focus on gender equality. In one of the projects they initiated in West Ghana, they worked to help women fishmongers organize. As is the case around the world, and throughout history, women are in the background. To use de Lauretis's term, they are in "the space off" of fisheries. They process and sell the fish that the men catch. Emelia Abaka Edu, a participant in WorldFish's training scheme, is clear about the gendered division of power: "It's the women who finance the men. We give them petrol, food, everything that they need to go to fishing. When they bring the fish, then we'll deduct the money that we used for those expenses and give the rest to them. So when they go and they don't get any fish, our money is lost" (WorldFish 2014).

Athwart

From mermaids to herring quines, from the collapse of the North Atlantic fisheries to the gendered relations of fishing in Africa, the importance of gender returns again and again. In this chapter, gender does a lot. Theoretically the ground shifts from my critique of how institutions and organizations have framed gender in a restrictive account of women's biological closeness to the environment to a more complex argument about the gender relatedness of humans and fish. In their review of what they call "ethnographies of encounter," Lieba Faier and Lisa Rofel argue that "ontologies of humanness are ongoing social processes involving encounters with other beings.... Becoming is always becoming with" (2014, 371). I don't think these feminists would mind if I pushed this into the gendered and queer ontology that Sedgwick gives us. This is to describe these ongoing social processes as athwart: as "continuing movement," as "recurrent, eddying, *troublant*" (Sedgwick 1993b, 12). To further add a methodological spin to Sedgwick's queering, Helmreich's use of "athwart" affords "an empirical itinerary of associations and relations, a travelogue which, to draw on the nautical meaning of *athwart*, moves sideways, tracing contingent, drifting and bobbing, real-time, and often unexpected connections of which social action is constituted, which mixes up things and their descriptions" (Helmreich 2009, 23).

From representations of mermaids to the tales of quines and the brave fish-women who faced down indifference and exclusion, gender has multivalence in this chapter. It is athwart history and experience. It propels and compels, it moves sideways and finds new force where eddying currents bring together other constellations of more-than-human woman and fish. Gender is the itch that stimulates "an urge to track down details and put together tales" (Tsing 2014, 237). That itch has led me to gather the stories related in this chapter. Gender in all of its relations requires that we hone "the arts of noticing the entwined relations of humans and other species across multiple non-nesting scales" (Tsing 2014, 233). There are devastating consequences of ignoring the multiscalar relations gender has had, and continues to have.

To return to Thompson's invaluable oral history, "Fisherwomen are exceptions, but they have a double importance. First, even as stereotypes, their existence helps us to see beyond the 'naturalness' of a male stereotype. Their own experience, their view of the world, would be well worth knowing and widely . . . illuminating. But second, these very exceptions are also a striking illustration of . . . the importance of *local variations* in the position of women, and of the historical roots of these variations" (1985, 7).

Here gender is not a thing and certainly not one thing. It is a constant force in bringing into being other worlds, and other ways of being-with.

This inspires a feminist method focused on interconnecting and mapping out a myriad of interrelations. It is perforce hesitant and modest, directed and tireless. This is to be keenly aware of Haraway's directions to be accountable, to always be on the lookout for the responsibilities we have to the many facets of the more-than-human—to feel this relatedness on one's skin. As one of the women in Davis's study says, "I just loves the fish"; "I just loves to touch the fish" (1986, 134).

5

Little Fish
EATING WITH THE OCEAN

> Why a buffalo song? Because the fish missed the buffalo. When the buffalo came to the lakes and rivers on hot summer days, they shed their tasty fat ticks for the fish to eat, and their dung drew other insects that the fish liked too. They wished the buffalo would come back.—Louise Erdrich, *The Round House*

In *The Round House*, Chippewa writer Louise Erdrich's novel, fish swim to the surface of the tales that the old man Mooshum recounts in his sleep. The complex structure of Erdrich's novel pushes at the limits of representation. The fish swim in and out, nibbling at the boundary between the extraordinary and the ordinary. In her deft treatment, we catch what happens when the world is seen as fluid, and where, as Stefan Helmreich writes, "things—refugees, nomads, weapons, drugs, fish—challenge borders because they are imagined to 'flow' across them" (2011, 137).

In this final chapter, I explore the small marine things with which we need to cultivate a closer relationship. These are the little fish and other marine organisms that disturb the classifications of what is edible, and for whom or what. In telling their stories, I want to disturb what we think counts as food. These are fish that are fed to fish, that become food for animals, and that are "reduced" to become health supplements for the wealthy. These little marine life forms include sardines, menhaden, anchovies, sea cucumbers, algae, and other entities low on the food chain, and some, like phytoplankton, too small to be seen by human eyes.

To use Eva Hayward's (2010, 580) term, maybe we need "fingeryeyes" (and fish fingers) to comprehend this liminal zone. As I argue, following little fish leads us to envision human and nonhuman relations within and across different trophic realms. In chapter 4, I ended with Tsing's call to practice "the arts of noticing the entwined relations of humans and other species across multiple non-nesting scales" (2014, 233). Here I want to extend her argument for "critical description." Tsing's anthropological project examines the intricate symbiosis between and among fungi, humans, trees, technology, and ethics. It is a project so interesting and smart that I grasp it at the edge of my brain. Her framing of "the entwined relations of humans and other species across multiple non-nesting scales" is deeply generative for my own project. How to eat the ocean well? How to eat with the ocean? Who eats what, what eats whom? Where? When? These are questions of scale that need to be very delicately handled.

In this chapter I engage the different histories and socialities forged between and among the little fish in the sea. When people ask me what we should eat from the ocean, the answer is always little fish—those that reproduce quickly, don't live too long, and humble and abase our gargantuan appetites for the disappearing top predators of the ocean. To little fish we can add bivalves, algae, and marine stuff that doesn't need a lot of other marine stuff to grow. But I don't want this to become a rigid morality that would result in dictates, causalities, and equivalences. This is about scalar intricacy and metabolic intimacies, not moral positioning. So instead of mounting a campaign for little fish against all else, here I continue with my relating, with my attempts to bring you into the worlds of fish relatedness: from sardines, to anchovies, to fish farming, understood not in a linear placing but rather as entwined multitrophic forms of which humans are but one. This is, as I wrote in the introduction, to begin in the middle of the ocean.

"The lovely animals of the sea, the sponges, tunicates, anemones, the stars and buttlestars, and sun stars, the bivalves, barnacles, the worms and shells, the fabulous and multiform little brothers, the living moving flowers of the sea, nudibranches and tectibranchs, the spiked and nobbed and needy urchins, the crabs and demi-crabs, the little dragons, the snapping shrimps, and ghost shrimps so transparent they hardly throw a shadow" (Steinbeck 1945, 29). John Steinbeck's description of intertidal dimensions from *Cannery Row* flows with the spirit of critical description. Steinbeck here channels his friend and collaborator Ed Ricketts, the marine biologist whose character and profession formed the basis of the character Doc—Steinbeck's very humane marine scientist. Steinbeck opens his short novel with this oft-cited line: "Cannery Row in Monterey in California is a poem, a stink, a grating noise, a quality of light . . . a nostalgia, a dream" (1945, 29). The stink was the smell of sardines, being canned and also mushed into fish meal. When the novel takes place in the 1930s, fourteen canneries lined Ocean View Avenue. Reporting on the history of Cannery Row, Renee Montagne says, "The view was a veritable ocean of sardines, pouring into the canneries. . . . In the 1930s and '40s, more than 100 sardine boats plied the waters off Monterey" (NPR 2003).

Steinbeck's poetic critical description, his art of nonjudgmental attention, relates the microconnections between and among all the elements in the more-than-human world that is Cannery Row. His approach resonates with the feminist approach that I have tried to elaborate, where the questions are modest yet crucial. As J. K. Gibson-Graham asks, "What can we learn from things that are on the ground?" (2011, 4). The task of the feminist here is not to judge good from bad. Rather it is to practice a willingness to take in the world. This is an "ethics of attunement, a more sensitive, experiential mode of assembling" (Gibson-Graham 2011, 4). Certainly I don't shy from condemning certain practices. But we need to be deeply contextual, particularistic, and modest in our engagements with that immensely unknown and unknowable entity we live with—the fragile oceans and their dependents.

"Fish She Is Very Small"

I love sardines. While I like them fresh and grilled, I have to admit that I especially love them canned. When I lived in Montreal, my favorite brand of sardines was the Club des Millionaires. I loved their slogan and the classy packaging with a fish in a top hat and tails (figure 5.1). And they were very good. They were also quite expensive for a sardine—not called the Club des Millionaires for nothing. These were sardines to aspire to, especially for a broke graduate student. The backstory of the brand is seductive. In the early

Fig. 5.1. Club des Millionaires sardines. http://www.millionairefish.com.

1900s the Club des Millionaires was a social club for rich businessmen in Montreal. I can see it in my mind's eye: They would have met in a grand parlor in one of the beautiful buildings on the Square Mile on Rue Sherbrooke, maybe trying to rival the famous Beaver Club that was set up for the entrepreneurs who made fortunes from beaver skins. Instead of trading in fur, they decided to start importing sardines. While it seems a strange direction for a social club, maybe in the depths of a snowy cold Montreal winter their thoughts turned to fish, or maybe they just loved sardines too. At first the businessmen imported canned Portuguese sardines. In 1908, Club des Millionaires Petit Sardines was launched, and it continues today. They switched from Portuguese to Norwegian brisling sardines because they are very, very petite.

The company takes its sardines seriously: "The brisling must be thronged for at least three days before being taken to the canneries. This process consists of pulling the catch close to shore and leaving them in the nets to digest the food already taken in. The swift, clear waters of the fjords take away the waste and the fish are netted so there is space to move, but they are prevented from feeding. This natural characteristic of the coastline—to allow thronging—can't be matched anywhere else in the world" (Harold T. Griffin Inc. 2015).

"Thronging"—what a wonderful word. Some parents might know of it from reading *Arlene Sardine* with their children. This wonderfully wacky book by Christopher Raschka opens with, "So you want to be a sardine. I knew a little fish once who wanted to be a sardine. Her name was Arlene." Arlene was a "happy little brisling because she had about ten hundred thousand friends." Arlene was caught in a purse seine net. She "swam around in the net for three days and three nights and did not eat anything, so her stomach would be empty. There is a word for this. The word is thronging" (Raschka 1998). While Arlene may have many friends, the idea of her three-day starvation is off-putting. I think I'll stick with the unthronged Australian sardines.

Sardines have been around humans (on and off, as we will see) for a very long time. They tell important tales of fish-human sociality. In the United States, canned sardines became popular in the late 1880s, and then they really came into their own during the period encompassing the two world wars. Cheap, nutritious, and easy to carry, they were the perfect war fodder. Before 1930, over 2,500 registered trademarks were in use in the sardine canning industry. When *Cannery Row* was published in 1945, the sardine industry was at its peak with 237,000 tons of sardines processed, although it was soon to crash—an ocean of little fish pouring into the canneries or flowing into the machines to crush and reduce them to fish oil and fish meal (Chiang 2008; Fotsch 2004). As Corby Krummer (2007) writes, "For decades in the 20th century [sardines'] abundance gave birth to an industry that fed millions of soldiers fighting both world wars and sustained thousands of Sicilians, Asians, and other foreign-born workers—the fishermen and packers of Cannery Row, in Monterey California—during the worst years of the Depression." Their association with the Depression may partially account for their downfall: "In the United Sates, sardines had always fought an association with food for the poor—the kind of thing you eat straight from the can in a cold-water flat." Krummer adds that sardines were definitively pushed off the shelf in the 1950s by canned tuna, which "became the wholesome food in cans, partly because it had no scary skin and bones."

As a foodstuff for humans, sardines have gone in and out of fashion. Recently Oprah Winfrey declared them one of her "top 25 super foods." But even the power of Oprah hasn't ensured them a secure place. In 2011, sardines returned to the British Columbia coast in schools "thick enough to walk on." The owner of one of the outfits making money from them admitted he didn't like them because they were "fishy." "They are exported to Russia, Ukraine and Asian countries where there is more of a culture around eating the oily fish" (Hamilton 2011). Fish that are fishy: Now there's a conundrum. And to remind you of why other food is fishy, please refer back to figure 1.1.

The process of reducing fish to meal or oil isn't pretty or straightforward. It has to be cooked, pressed, dried, and ground, and along the way it produces a powerful smell of rotting fish. It also uses a lot of energy as modern-day factories dry the meal at five hundred degrees Celsius. "Reduction fisheries" are also called industrial fisheries, and both designate an operation devoted to grinding and cooking the fish—none of which goes for "direct human consumption." The rather gruesome appellation matches an inequitable state of affairs. Roughly a quarter of the global fish capture is for "nonfood" (read not food for humans). To put this figure in other terms, WWF (2015b) argues that "seven of the world's top ten fisheries (by volume) target forage—also known as low trophic level—fish, 90 percent of which are processed into fishmeal and fish oil." Peru has the largest single-species fishery in the world, and less than 1 percent goes to direct human consumption. The species is the little Peruvian anchoveta, to which I will return. It provides some 30 percent of global fish meal and oil.

Fish in a Bottle: The Battle for Omega-3

I'm looking at a bottle of fish oil capsules that could have been sourced from any number of websites, branded by any number of producers. This one happens to be Australian, Blackmores, one of the more expensive brands you can find in any Australian supermarket or chemist. Playing on the "pristine" nature of Australia, it is now big in the Chinese market for non-Chinese food supplements. Blackmores, like all health supplement companies, is "passionate about natural health and inspiring people to take control of and invest in their wellbeing." This particular fish oil is marketed as "anti-inflammatory," while others inspire us to take control of our "optimal health." On their web page the bottle stands on an attractive kitchen table that I would peg as vaguely Scandinavian. Trendy triangle silk teabags nudge the wooden bowls. It screams middle-class taste. Blackmores has joined with WWF, so on their website they now have the WWF

Fig. 5.2. Tip Top omega-3-enriched bread. Photograph by author.

black-and-white panda logo with the message "Working with Blackmores for sustainably sourced fish oils" (WWF 2015a). To conduct a critique of the neoliberal incitement to self-fashioning governmentality (from furniture to fish oil) that this image evokes would be like shooting fish in a barrel.

Among the myriad of its advertised benefits (from child development to a treatment for Alzheimer's prevention, and everything in between), omega-3 is apparently good for thinking, and maybe good for thinking with. In Australia, bread now comes fortified with omega-3. Tip Top The One bread is not the cheapest (figure 5.2). It has a long history of being positioned as the quality bread mothers feed their kids—the slogan for years was "Good on ya Mum." However, it sells for about three dollars versus the nearly thirty dollars for the Blackmores fish oil capsules. Wealthy mothers can buy differentially branded omega-3 fish oil capsules for themselves, their children, and even for their pets. However, working-class mothers may be at best able to afford omega-3-enhanced bread. Mothers deluged by media factoids about omega-3 aspiring to feed their children "special" bread is an image of such cruel optimism (Berlant 2011).

Thinking with omega-3 informs Jane Bennett's (2009, 39) theory of "vibrant materiality," where she takes the various claims about omega-3 as a springboard into her theorizing about food as an actant. Bennett "finds support in scientific studies of the effects of dietary fat on human moods and cognitive dispositions" (2009, 39–40). These studies demonstrate that the marine omega-3 fatty acids eicosapentaenoic acid (EPA) and docosahexaenoic acid (DHA) "can make prisoners less prone to violent acts, inattentive children better able to focus, and bipolar persons less depressed" (41). Here I want to extend Bennett's take on omega-3 fatty acids into what its production does to the ecologies of humans and fish. From the perspective of science and technology studies, Annemarie Mol's Eating Bodies in Practice and Theory team takes issue with how Bennett treats fish oil and omega-3 in isolation. They are also critical of the desire to find agency at all costs. They conclude, "The case of omega-3 helps to explore other modes of doing, such as affording, responding, caring, tinkering, and eating" (Abrahamsson et al. 2015, 6). In their lovely phrase, "Matters may engage in relations of ever so many kinds" (6). Abrahamsson et al. construct a three-pronged argument, the first of which is "what omega-3 may do." The second is the question: "Where does omega-3 come from?" (11). "If omega-3 is sourced from fish for human consumption, this may, in the short term, improve our moods, but in the slightly longer term it depletes the oceans. In the process not all human beings are affected equally. . . . Omega-3 is not so much 'matter itself' as 'matter related'" (12).

Of course one can source omega-3 from plants and from eggs. There is some consternation about which type of egg is best and what defines a better egg. Free-range pasture-fed eggs are said to be the best bet for omega-3 (the ones fortified with omega-3 are the highest, although that rather obviates the point of bypassing fish). In Australia the definition of free-range is pretty loose, and currently we don't have a legal definition of what constitutes free-range. In common sense, one would think that if chickens eat what they seemingly like to eat—green stuff, insects, and worms—they will be healthier and maybe even happier (Miele 2011). As the website Authority Nutrition tells us, "This just goes to show that what we eat isn't all that matters. . . . It also matters what our foods eat" (Authority Nutrition 2015).

Free-range, grass- and insect-munching chickens apparently do produce eggs higher in omega-3, but they pale in comparison to sardines. Three ounces of canned sardines gets you 61 percent of your Dietary Reference Intake (DRI) of omega-3, whereas an egg nets only 3 percent (George Mateljan Foundation 2015).[1] The overwhelming majority of omega-3 oil capsules currently come from fish (or krill, a zooplankton crustacean rich in phospholipids carrying EPA and DHA). But as Abrahamsson et al. write, "this fish, in its turn, is not 'fish itself,' but 'fish-related' too" (2015, 12). In the case of fish oil, this is why fish contain omega-3 in the first place—because the little ones eat phytoplankton that generates what becomes omega-3. They are in turn eaten by the bigger fish. "Fish-related" thus encompasses the fish's relationship to its habitat—from the macro scale of the ocean down to its local practices, whether it is pelagic, benthic, or demersal (i.e., where it hangs out in the water column). The fish that are "reduced" (in every way) to fish oil and fish meal are pelagic (from the Greek *pélagos*, meaning "open sea").

Little Fish Lines

As Abrahamsson et al. point out, the matter of fish oil is of course fish themselves and what they ingest. They cite one of the few studies that examines what fish oil does to human and fish ecologies: where it comes from and with what consequences. The title of the article puts it succinctly: "Fish, Human Health and Marine Ecosystem Health." The subtitle says it all: "Policies in Collision" (Brunner et al. 2009). Coauthored by a team of epidemiologists and a geographer, it homes in on the risks and benefits of fish oil—on the one hand the public perception is that fish and fish oil are good for us, but on the other the normal intake isn't at the much higher recommended levels. Then for some—pregnant women—even a fairly small intake of oily fish high up the trophic ladder can lead to methylmercury exposure that might harm the fetus (Brunner et al. 2009, 94). The sad reality is that large predator fish such as tuna or swordfish have been exposed to high amounts of methylmercury indirectly through fossil fuels and coal-fired forms of energy. Just as in humans, it accumulates in their fat and cannot be metabolized. The more little fish the big ones eat, the more concentrated the amount of methylmercury becomes, and as humans tend to want to eat the big fish, we ingest it too. In Becky Mansfield's analysis, environmental contamination is "metabolized by individual fish, such that they become a vector for moving the contaminant into human bodies" (2010, 424). It's a conundrum. On the one hand people in places like the United Kingdom and the United States don't really eat enough oily fish to get the vaunted omega-3 benefits, and yet for pregnant women a little can be too much. As Brunner et al. argue, this produces a policy quandary: "the problem of balancing risk communication in populations that in general consume little oily fish" (2009, 94). It's a case where humans end up eating and getting sick from what we've made the fish consume.

So instead of eating the top predators that contain the most toxins, we are encouraged to get our fish from a pill. With the heavy resources of Big Pharma and media and advertising to persuade us to take fish oil capsules, there has been a large increase in their consumption. Ten percent of Americans take fish oil capsules, making them second only to vitamin supplements (Berkeley Wellness 2014). Because the capsules are made from fish low on the trophic range, they contain much less toxin than bigger and older fish. There has been a concomitant effect on fish stocks, as well as the people who rely on fishing them. Desperate to get foreign currency and often at the behest of the World Bank, developing coastal countries effectively sell off their national Exclusive Economic Zone rights to natural resources to foreign supertrawlers. Brunner et al. write that "low cost access agreements with West African nations such as Senegal and The Gambia, where fish provides over 50% of dietary protein, permits EU trawlers to take tens of thousands of tons of prawns, tuna and hake and other high-value fish annually for European consumers" (2009, 97). In Abrahamsson et al.'s framing: "As human eaters organize themselves in complex sociomaterial ways, the fish they eat has become entangled with long-distance routes" (2015, 9). These are simultaneously legal, technological, political, and humanitarian issues that frame who has access to what and where. In their riposte to Bennett, they write, "Rather than getting enthusiastic about the liveliness of 'matter itself,' it might be more relevant to face the complexities, frictions, intractabilities, and conundrums of 'matter in relation'" (Abrahamsson et al. 2015, 10).

I want to take their framing of the "matter of relation" farther out to sea. One small matter that the Brunner article doesn't delve into is that the vast majority of fish caught up in the reduction industry are not the high-value ones they cite: prawns, tuna, and hake. It's the fish called trash, defined as unfit for human consumption—defined, that is, by the International Fishmeal and Fish Oil Association (IFFO), widely known as the OPEC of fish oil. Peru, Chile, and Denmark are the leading producers of fish oil, accounting for over 50 percent of production (Transparency Market Research 2015).

"Industrial grade forage fish" (the lowest on IFFO's ranking) are members of the Clupeidae family, including sardines, herrings, menhaden, and anchovies. Along the Atlantic coast of North America, the little menhaden once reigned. These are small silvery fish with spots. H. Bruce Franklin (2007) calls them "the most important fish in the sea." Franklin's book traces what the menhaden do and have done in the United States—as his subtitle has it: *Menhaden and America*. From long before European contact, these little bony fish fed the crops of First Nations peoples (they planted corn with crushed menhaden). The colonizers soon scooped the fish up. By the 1870s, "menhaden had replaced whales as a principal source of industrial lubricant, with hundreds of ships and dozens of factories along the eastern seaboard working feverishly to produce fish oil" (Franklin 2007). They have been made into lipstick and soap. Omega Protein Corporation has captured them since 1913, and the company now has a monopoly on the menhaden fishery in the United States. Franklin's groundbreaking research connects the science and the history of the menhaden. Their future isn't looking good, nor is that of the entire ecosystem of the northeast Atlantic. As in Mansfield's (2010) argument about the recursive relationship between environment and commodity, the removal of menhaden from their ecosystem is unleashing new and toxic relationships: "The vital biolink in a food chain that extends from tiny plankton to the dinner tables of many Americans appears to be threatened. [Due to a lack of menhaden in their diet] . . . half the rockfish in the Chesapeake are diseased, with either bacterial infections or lesions associated with *Pfiesteria*, a toxic form of phytoplankton known as the cell from hell" (Tavee and Franklin 2001).

None of this seems to bother Omega Protein. Their website shows a smiling, tanned, blond all-American family in their tasteful kitchen, and we read, "Omega Protein is the leading, vertically integrated producer in the United States of omega-3 rich fish oil, protein-rich specialty fish meal and organic fish solubles for livestock and aquaculture feed manufacturers. These vital feed nutrients are passed on to the human consumer in the form of sustainable, farm-raised seafood and pork. . . . Omega Protein provides companion animal manufacturers the right ingredients to improve protein quality and provide additional nutrients like omega-3 fatty acids to help improve health and performance" (Omega Protein 2015).

The text moves smoothly over the immensity of what happens to the little fish in order to pass through to the human consumer via pigs, chickens, farmed fish, and fish oil capsules. Reduced to their "vital nutrients," the menhaden seep into other spheres, becoming the basis for industrial pig feed, pet food, and of course fish farming. As Paul Greenberg (2009) notes, "Menhaden filter-feed nearly exclusively on algae, the most abundant forage in the world, and are prolifically good at converting that algae into omega-3 fatty acids and other important proteins and oils. Starkly put, they form the basis of the US Atlantic Coast's marine food chain." In addition to providing "vital nutrients" for Omega Protein's products, they are also essential to the marine ecosystem. In Greenberg's words, "Nearly every fish a fish eater likes to eat eats menhaden." As a filter feeder, menhaden eats only phytoplankton. This part of the menhaden's nature is crucial: "An adult menhaden can rid four to six gallons of water of algae in a minute. Imagine then the water-cleaning capacity of the half-billion menhaden we 'reduce' into oil every year" (Greenberg 2009). Like the oyster, if left to do what it naturally does, menhaden can rectify the quality of seawater, turning brown sludge into clean and clear seas. That it eats phytoplankton—microscopic bits of algae—is crucial in two ways. First, the phytoplankton gives fish its omega-3. If the bigger fish don't eat the little fish, they begin to be depleted of omega-3. Second, the menhaden, like other forage fish, is very much a fish-related fish, highly connected to many other species across the food web. While phytoplankton is essential, too much of it causes hypoxia. This is what causes the massive algae blooms. If allowed to, the menhaden can eat into the cause of oceanic dead zones where nothing except jellyfish can live.

La Rica Anchoveta

The beautiful fish in figure 5.3 is the Peruvian anchoveta (*Engraulis ringens*), which is often described as *la rica anchoveta*, the rich anchovy—rich in omega-3 and also the basis of past and present wealth in Peru. It is a dark-blue, greeny-silver pelagic schooling fish and measures about four inches long. Swirling glorious balls of anchovies are a favorite for underwater photographers when predators attack and the fish form a massive bait ball. The

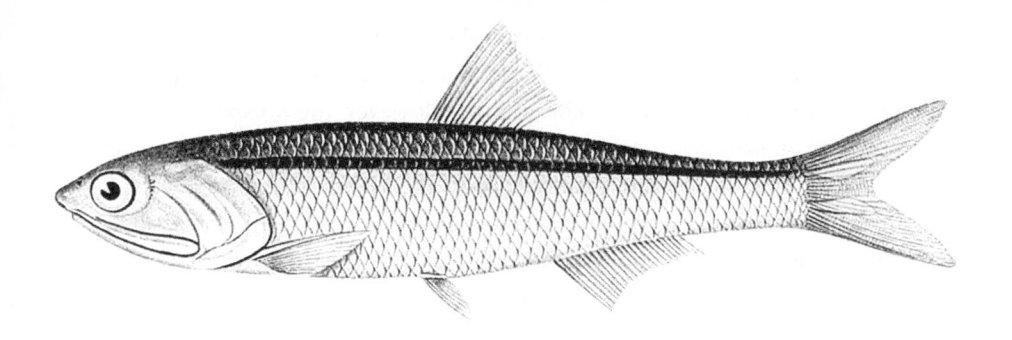

Fig. 5.3. Anchoveta peruna (*Engraulis ringens*). https://de.wikipedia.org/wiki/Engraulinae.

anchoveta is classified as food-grade forage—one step up from the menhaden. According to IFFO, 486,500 tons of Atlantic menhaden were landed in 2011. By comparison 8,468,000 tons of Peruvian anchovies were caught. This little fish is the world's largest fishery by far. Since 1960, this single species has accounted for about 10 percent of the total world finfish catch. The reason why the Peruvian waters are so fecund is that the Humboldt Current (the Peru Current) brings to the surface the highly nutrient-dense deep waters, and unusually this upwelling is relatively close to shore. Many claim that Peru's civilization extends back before the Incas, and it was based on the anchovy (Pauly and Tsukayama 1987; Coutts, Chu, and Krigbaum 2011). In 1908, Dr. Robert E. Coker, the fishery expert to the government of Peru, told the Fourth International Fishery Congress in Washington about "the immense schools of small fishes, the 'anchobetas' (*Engraulis ringens Jenyns*), which are followed by numbers of bonitos and other fishes and by sea lions, while at the same time they are preyed upon by the flocks of cormorants, pelicans, gannets, and other abundant sea birds" (Christensen et al. 2014, 302).

In 1948, General Manuel A. Odría Amoretti staged a military coup to become president of Peru. His plan for economic development was based on exploiting the country's natural resources. With the collapse of the Monterey sardine industry, entrepreneurs bought cut-price boats and fish meal equipment and moved their operations south to Peru. This coincided with the huge expansion of the American pig and chicken industries to provide cheap protein in the post–World War II period. In turn this spurred demand for fish to feed the cheap meat. However, the prosperity of the new fish oil manufacturers had to contend with another more-than-human element: guano—the nitrogen-filled poop of cormorants and boobies that feed on anchovies. The Peruvian guano industry dates back to what is called "the Guano Era" in the mid-nineteenth century, when the worldwide demand for the highly nutrient-rich seabird dung was at its peak. Guano was the favored form of fertilizer. The birds needed to be able to feed on the anchovies, which were being depleted by the fish oil factories. The ecosystem needs them both.

In the early 1950s, the guano industry was still dominated by the wealthy Peruvian elite. They did all they could to stop the fishery. The first anchovy plant in Peru had to be set up in secret, and in 1956, thanks to the guano lobby, the government halted all construction of plants. However, in 1959, realizing how much the fish were worth compared to the bird manure, the government lifted the moratorium. The fishery and the reduction factories boomed at the expense of the seabirds. By the 1970s, some 14,500 purse seiners were trawling the waters with the capacity to catch thirteen million tons of anchovies, despite the quota being set at 7.5 million. And there were enough fish plants to process eight thousand tons a day—far outstripping the legal catch (Lewis 2000). As the comprehensive report *Little Fish, Big Impact* states, "The direct exploitation of anchoveta made the ecosystem less resilient to El Niño events, and the first of these after the onset of this fishery in 1965 saw the massive decline of the huge seabird populations which had until then maintained the large Peruvian guano industry" (Pikitch et al. 2012, 48). In 1972, a socialist military junta seized power and nationalized the industry. But a powerful El Niño devastated the anchovy fishery, and the government privatized it.

The combination and interrelation of events is extraordinary. The little anchovy is embroiled in a tale of epic proportions. To consider just some of the elements, the collapse of the sardine fishery in California allows cheap fishing gear and fish meal technologies to move down to Peru. The anchovy contends with El Niños, which sometimes deliver pockets of cold water in which it can flourish but more often than not bring warm water, which favors sardines rather than anchovies. This mighty fish is crucial in the balance of more-than-human ecologies.

Eat Little Fish: La Semana de la Anchoveta

What if humans ate anchovies before they were pulped to feed other animals and fish? This seemingly obvious idea was the brainchild of Patricia Majluf. Given how economically imbricated the anchovy is in Peru, it was hardly a straightforward proposition. Majluf is the director of the Centre for Environmental Sustainability at Cayetano Heredia University, and she was trained as a zoologist at Cambridge University. Her area of specialty was seals and sea lions. After a major El Niño, her research subjects disappeared, and she turned to little fish.

I first met Majluf at the Slow Fish Genoa event in 2013. Slow Fish is the piscine equivalent of Slow Food, focusing on artisanal fisheries. It's a daunting event—tens of thousands of visitors and hundreds of exhibitors. I made my way through the crowds with difficulty, pausing occasionally to taste a local tidbit. I arrived at the tent where the session "Can Fish Feed Fish?" was being held and immediately recognized Majluf from her television and film interviews. Silvio Greco, a marine biologist and the president of the Slow Fish Scientific Committee, started by asking, "Is it ethical to catch fish to feed fish?" Greco reminded us that the Peruvian anchovy fishery was destroyed in the 1970s "to feed chicken that tasted like fish." Next up was Alain Le Sann of the Collectif Pêche et Développement in France. Le Sann is a passionate advocate of artisanal fisheries. Having worked in NGOs in Africa, when he retired to his birthplace in Brittany he set up Le Festival de Films Pêcheurs du Monde—the first, and only, international film festival devoted completely to films about fishers around the world. In his talk, Le Sann spoke about the plight of the fisheries, the fishers, the fish, and the people of Mauritania. Dozens of industrial processing plants for fish meal are being set up by the Chinese in Mauritania, processing fish into fish meal to be sent to Chinese fish farms that could have been eaten by the local population. Because of the high price of fish meal, fish has become more expensive than meat in some parts of the Global South where previously fish was the cheap mainstay of the diet. The painful irony is that the fish meal reduction industry creates very few jobs and generates little income for these poor fishing communities. Large trawlers scoop up the fish, and without any refrigeration they are left in the holds to putrefy before being pumped into the fish meal plants.

Le Sann's talk set the stage for Majluf. She started by drawing attention to the sheer size of the anchoveta fishery. Peru is a relatively small country with a high socioeconomic and class divide between rich and poor—the GDP is just over eleven thousand dollars per capita. The national rate of chronic malnutrition for children under age five is 18 percent, and in the rural Indigenous areas in the south it can be as high as 43 percent (Páez 2011). And yet, according to Majluf, half a kilo of anchovies a week would provide complete protein needs. Instead the children are fed potatoes. The problem is that, as in Mauritania, huge amounts of anchovies go straight to the reduction factories and then onto boats to feed the fish in China's fish farms. Kristin Wintersteen describes the problems caused by this global food chain: "[It] relies on the large-scale extraction, processing, and export of fish meal from the global South to industrial livestock and fish farms, relegating the environmental costs (overfishing, air pollution and water pollution) to the people and ocean of Peru. In this system, nutrients are removed from the marine ecosystem and the local food chain is then transferred *en masse* to distant markets and, eventually, distributed to consumers who are disconnected in every way from the socioenvironmental conditions of its production" (2012, 628).

Majluf was determined to intervene and resolve the interrelated problems of malnutrition, pollution, poorly paid jobs, and overfishing by getting people to eat anchovies. First she needed some anchovies, no small feat when they all go to the reduction plants. "I asked where I could get fresh anchoveta. I was put in touch with Juan Bacigalupo, the only one who sells fresh anchoveta in Lima. I tried them. They were delicious, better than the best Spanish sardine. He gave me a few cans" (quoted in Wintersteen 2012, 630).

Next was the problem of how to persuade people to eat them. Lima is now widely recognized as one of the food (and foodie) capitals of the Americas. Gaston Acurio has spearheaded this effervescence of gastronomy. The French-trained chef returned home to Peru, and from Lima he has taken the world by storm. He is called "a culinary ambassador with dozens of Peruvian restaurants around the world—Astrid y Gastón in Lima made S. Pellegrino's 2013 list of the top 50 restaurants in the world." As the

LA Times puts it, "The twinkle-eyed Acurio has marched decisively forward as the commander-in-chief of Peru's gastronomic revolution, a process of discovery and dissemination of native flavors that he hopes will captivate the world" (Barclay 2009).

So to whom did Majluf turn? "I was able to get an appointment with Gaston—the only one who could get people to eat such a thing. I went with my cans and asked him to try it. He did, he said how delicious, what can we do?" (quoted in Wintersteen 2012, 630).

Late in 2013, I invited Majluf to Sydney for the Marine and Maritime Research Festival we organized with fishery scientists, oyster growers, fishers, marine biologists, cultural studies scholars, political scientists, and historians.[2] Such diversity is the norm for Majluf. She regularly consults with high-powered fishery and government officials around the world—indeed, for a short period of time she was the vice minister for fisheries in Peru.

Majluf's (2013) title for her talk was "The Very Elusive Win-Win-Win (A Story of Greed, Overfishing, Perceptions, Luck, Peculiar Circumstances, and Hopefully a Happy Ending)." To get an anchovy on the table, she had to work on several dimensions: "Working with an international network of producers, environmentalists, fishmongers and chefs, the plan [was] to produce a premium product that will change public sentiment and rejuvenate small-scale fisheries in Peru. Processing is key; the fish are salted and put in a barrel to cure for six months. The result is far from malodorous. 'They look and taste more like sardines, and a lot of people say when you open the can no one can tell they're anchovies,' [Majluf] said" (Cuthbert 2014).

Majluf's three wins in her title refer to the three greatest obstacles she faced. First, the big business of fish reduction: IFFO represents all the fish meal and fish oil processors in the world. That's a lot of power and a lot of potential for corruption. As I flagged earlier, IFFO came up with the designation of the fish they reduce as inedible—or not for "direct human consumption." On their website, they argue that "only 10 percent of industrial fish has a market for human consumption" (IFFO 2011). But again, this is because they have already labeled the fish as "industrial" and because they make no attempt to treat the fish as anything other than industrial fodder. As Majluf was to find, IFFO had a pretty direct line into the Peruvian government. Until recently, all the ministers in the Fisheries Ministry had come from the reduction industry, and corruption was rife. When she resigned as vice minister of fisheries, Majluf told reporters, "I entered a ministry that traditionally has been controlled by the industry, so it has lots of processes that are designed not to work." When asked to elaborate, she did:

> The fishing licenses, the fines. Of all the infractions that occur in Peruvian waters, less than 10% are punished, and of that 10%, no one pays. . . . The boats enter within 5 miles (of the coast) or catch fish of prohibited sizes, the GPS don't work, the scales lie, false holds, blocking the inspectors' access, licenses to fish sardine while actually fishing for anchovy, licenses to fish anchovy for human consumption but actually using it for fishmeal. There are tricks for every step. When I made a presentation and said, after investigating, that there was an undeclared catch of 20–40% above the limit for anchovy, they yelled and screamed because it was admitting that we were overfishing. (Labarte 2012)

Unfortunately, many fisheries around the world are renowned for being liable to corruption. In Sicily, the illegal tuna trade is associated with the Mafia. In Thailand, some fishing boats use Indonesian slave labor. Often governments react against scientific advice, sometimes because of the lobbying power of fisheries, and sometimes in reaction to public outrage, often galvanized by powerful lobbies.

Majluf's win against the powerful fishery, the ministry, and IFFO was no small feat. She has convinced the fishers that selling anchoveta as food for people, rather than as fertilizer or animal feed, can be profitable. By value adding the catch, you can reduce the total allowable catch without affecting the industry's income. She got the individual transferable quotas changed to allow artisanal fishers to be able to catch anchovies for human consumption. This encouraged small boats to invest in the necessary equipment to keep the anchovies fresh. As a result, the human consumption of anchoveta in Peru rose from 10,000 tons in 2006 to 190,000 tons in 2010.

The second challenge was to change people's ideas about the little fish. She needed to "reposition" anchovies "from food for animals to trendy chefs, to make it food you aspire to" (Majluf 2013). This was kick-started with Acurio, but perhaps as important was to make the product look good. From the outset, her team ("a cook, a biologist, a designer") wanted to appeal "to people's stomachs and eyes" (Majluf 2013). Majluf and her colleagues organized huge media events to broadcast the idea of eating anchovies. She persuaded the then president of Peru, Alan García, and his whole cabinet to eat anchovies on television. And they liked them—as did 95 percent of the people who ate them. They set up La Semana de la Anchoveta (which has happened annually since 2009). At the first one, they had a hundred restaurants on board and ten thousand people tried anchovies for the first time. They printed beautifully designed cards with the salient facts about the health aspects and about the fish and its ecosystem. Within three months, the president had signed a "law affirming the strategic importance of promoting the consumption of anchoveta and its nutritional properties among Peruvians." They put eighty million U.S. dollars, or 30 percent of the food security budget, toward producing and promoting anchovies. They featured "anchoveta breakfasts" in schools (Wintersteen 2012, 631). Majluf convinced them that "it was all their idea."

Majluf and her team wanted food for the poor to taste good and to look good. This is so obvious and yet so radical. They froze the anchovies in packages that could be easily transported to the Highlands and the regions where the poorest live. They got them into schools and hospitals. But they didn't want to stigmatize anchovies as a food for the poor. The middle classes were encouraged to buy into the anchovy bandwagon via the arguments about the health benefits of omega-3 ("More Anchovies Please" 2011).

As Wintersteen frames it, this is a "market-based strategy for improving environmental sustainability in the Peruvian anchoveta fishery" (2012, 626). And indeed, economics hand in hand with good design and good taste meant that companies began to invest in processed value-added products (with over fifty-six different anchoveta products in local markets).

Before she left Sydney, Majluf came over to our place to eat oysters and wild prawns from our bit of the Pacific. Eating the prawns, heads and all, she talked about what was next. "Anchovetas for the world!" She had just been awarded Pew Trust funding for a project titled Harnessing Market Forces to Shift the Largest Reduction Fishery on Earth toward Sustainability. This will involve scaling up the already pretty impressive anchoveta campaign to work with an international team of lawyers funded by Pew to figure out how to harness market forces and change regulations at global and local levels. They want to come up with new formats, engage with big retail chains to "make it easy, cheap and tasty." She also wants to draw out more goodness from the anchovy by using the waste for fish sauce, fish salt, and anchovy paste. And why not? In the West, anchovies have been used since the Roman Empire to spice up food. I remember Patum Peperium

Fig. 5.4. Anchovetas for the World poster produced by Centro para la Sostenibilidad Ambiental, Universidad Peruana Cayetano Heredia. Photograph by author.

Gentleman's Relish, an anchovy paste that dates back to the 1820s. It was only for special occasions, when it would be thinly spread on buttered toast with watercress.

Metabolic Intimacies

Majluf's project relates anchovies and people, working with the anchoveta's deeply fish-related materiality. The profound relatedness of fish and human reminds me of Annemarie Mol and John Law's take on more-than-human "metabolic intimacy" (Law and Mol 2008). As a concept, it directs us to think about the multiple trophic and structural levels through which we (fish, humans, animals) are related. In their article, Law and Mol follow an outbreak of foot-and-mouth in a pig farm in the north of England. It's an intricate tale, but the main points are that the disease—which is not harmful to humans—was reactivated in the United Kingdom because the farm had not been boiling their pig swill. Pigs traditionally are fed with human leftovers, in this case from a cafeteria. But the farmers hadn't boiled the pig swill, which kills the virus. As Law and Mol define this technology, "So this is boiling pigswill. It makes boundaries: between untreated and treated waste food; and between pigs on the one hand and bacteria and viruses (the foot and mouth virus is just one of these) on the other" (2008, 135).

Two important things happened as a result of the farmers' laziness and disregard for their pigs: "As a result of the foot and mouth outbreak, feeding pigs with swill from catering waste was made illegal in the UK on 24th May 2001. This put an end to an English history of human–pig intimacy—let us call this a metabolic intimacy that goes back at least 500 years" (Law and Mol 2008, 137). Not only is the connection between pig and human broken, but British pigs are now fed with soy shipped from Argentina that "would have been perfectly suitable to feed the rural population of Argentina—or elsewhere" (140). They conclude that "boiling pigswill is—was—a political technique for avoiding waste. It drew human food that was thrown out of it, back into human metabolic circuits. . . . Boiling pigswill was a political technique that, in a region of plenty, respected and helped to limit food scarcity on a worldwide scale" (141).

What I particularly like about this analysis is how small, seemingly incidental events are drawn into a global web of relations and relatedness. When human error breaks up centuries of human-pig intimacy, soy is remade into a monocrop for animals, fed for the delight of certain privileged humans. This tale exemplifies the destruction of diversity at different scales: of the biodiversity of crops in Argentina and of the diversity of more-than-human relations of pig-human metabolic intimacy. While pig swill is of a different order than the little fish, the traffic in metabolic intimacy at global and local levels is analogous. Through the technology of reduction, exported from California after humans had wiped out the sardines, metabolically the fish are connected to pigs and poultry and other fish, and then in a very diluted way to human eaters. Instead of feeding humans, the anchovies are reduced (again in every way) to stuff sent across the world. It's not that this stuff isn't still vital, to recall Bennett's terms; it's just that it is deanimated, and its relatedness to several orders is curtailed.

Tinkering with Multitrophics

And what of the fish farms, the most important end point of the reduction industry? Despite being reviled by many, and for many reasons, aquaculture is the fastest-growing global food production system. It can be seen as a capitalist fix to the crisis of capture fisheries, as Mansfield argues. In this sense, aquaculture threatens to make "the crisis worse, because fish overall are becoming cheaper just as wild fish are becoming more scarce and expensive to produce" (Mansfield 2010, 423). But equally, aquaculture is poised to answer the question of whether the "oceans [will] help feed humanity" (Duarte et al. 2009).

A ten-year UNESCO-funded study of the structure and functioning of the global ocean ecosystem, GLOBEC, argues against "pitting aquaculture against fisheries [as] both activities urgently need further research for integrated management and sustainable development" (Freon et al. 2010, 28). The scale of aquaculture makes it hard to be simply for or against: "Taking place in approximately 190 countries and involving cultivation of roughly 600 species—from salmon to oysters to sea urchins—aquaculture supplies more than half of all seafood produced for human consumption" (Future of Fish 2014, 1). In addition to the hundreds of species being farmed, there are millions who depend on them. Meryl Williams of GenderAquaFish (an NGO-supported center of research into gender and aquaculture) writes, "Aquaculture and fisheries are very diverse, and so are the societies and gender norms in which they operate" (Gender in Aquaculture and Fisheries 2014). Her point is crucial, as is the recognition that for many women and their families around the world, aquaculture is often the only means of food subsistence as well as income. As we know, there is a very real crisis in land-based food production in many parts of the world. And yet the view persists that aquaculture only consists of huge salmon farms for which little fish are ground up and squandered. Ironically, this type of reaction is common in Australia, a land that is now so water depleted, hot, and salinated that the future of land-based crops and grazing is seriously in question. And yet the various levels of government involved have tightly restricted aquaculture, with few new licenses allowed in the past ten years. Despite several healthy and MSC-accredited seafood farms in Australia, the public perception (fed by sensational media reports and perhaps a dose of food racism) is that aquaculture is always and only a destructive practice that happens in places like Thailand or Vietnam (SBS 2014).

There are, of course, reasons to worry about aquaculture, including water pollution, concerns about farmed species interbreeding with wild ones, welfare concerns, and the very ethics of feeding fish to fish. As Mansfield (2010) argues, the production of a different and denatured fish—North Atlantic salmon do not by nature belong in huge pens in the Pacific—is bound to produce new recursive relationships. For others such as Patricia Majluf, eating farmed salmon as it is currently produced is a waste of anchovies. For many, the environmental concerns are the most pressing issue: the farmed fish poop problem. Fish fecal matter can cause serious health concerns for humans. An oft-repeated line is that a Scottish salmon farm produces as much fecal matter as a town of 65,000 (Pure Salmon Campaign 2015). This also contributes to nitrogen overload, which as I described earlier eats the oxygen out of the water.

The concept and practices of fish farming go back millennia. In the West, oyster cultivation and fishponds were a mainstay of upper-class Roman life across their large empire. Lucius Junius Moderatus Columella, renowned as the most important Roman authority on agriculture, recommended that seaweed be placed in the fishponds, "as far as the wit of man can contrive, to represent the appearance of the sea, so that, though they are prisoners, the fish may feel their captivity as little as possible" (Grout 2015). And as we will see, in Asia the ancient practices of polyculture form the current basis of a massive industry.

Large-scale aquaculture is a very recent development, and practices are changing—perhaps faster than perceptions. In the west of Scotland where the majority of the Scottish salmon farms are sited, I talked to the sustainability officers for one of the largest salmon-farming companies. Penny and Matt are part of the new breed of people involved in salmon farming. Young and politically aware, Penny has a background in geography and worked in marine conservation for several years before joining the company. She clearly still sees herself as a conservationist and volunteers on Ocean Watch's beach-cleaning projects. She speaks of the challenges of revitalizing fish farming in this part of the world where it's only less than sixty years since it was first implemented. This means "across one generation, people can remember the bad practices like when people just chucked bags of food into the pens." They are keenly aware of the social and environmental tensions in these small communities. On the one hand there are few prospects for employment, especially for the young, because there is so little industry in rural Scotland. On the other hand the beautiful landscape and proximity to Glasgow mean that there are a lot of holiday second homes. Penny and Matt describe the situation: "The born-and-bred locals are pro–fish farming," whereas the retirees and the holiday homeowners are "less accepting." In addition to not wanting their views spoiled by the circles of pens in the lochs, these middle-class owners of holiday homes would join with the anti–fish farming brigade, while the locals worry about getting work.

Matt and Penny's job is to intervene in this "balancing act between local employment versus the fact there's little space left in the West Coast for more farms." There is a lot of tinkering on the ground. For instance, when one of the locals asks them why they couldn't have a community salmon cage, the answer is, "Why not?" In these small communities, there's not a lot of room for blanket condemnation. The approach needs to be about what a place can afford and open up rather than close down. This type of tinkering takes place within an intense web of new local, regional, and global regulations—concerns about the welfare of fish are now front and center as business and environmental concerns converge. Tastes also change. Increasingly, few Highlanders would remember the taste of wild salmon (unless it's illegally caught). The fact is that if they want to eat salmon, it is farmed. It's the same around most of the world, with the exception of some places such as the Pacific Northwest of North America. However, wild Pacific salmon are in peril, largely because of the dams that stop them from homing to where they breed.

At a larger scale, the tinkering with farming fish becomes ever more high concept as it brings together the past and the potential. Thierry Chopin, one of the world's leading aquaculture scientists, is clear that "the aquaculture revolution, the blue revolution, has not always been green. We have to make the blue revolution greener. We need a turquoise revolution!" (Future of Fish 2014, x). Chopin's work builds on aquatic polyculture practices that have endured for millennia in many Asian countries. This was, and still is in remote areas, a perfect little trophic system: A duck swims in a rice paddy and poops, thereby fertilizing the rice. Some of it trickles down to the carp that are eating the vegetation in the paddy, and who also scoop up the detritus in the feces. Add some vegetables grown nearby fertilized by remains of the carp, and irrigated by the pond water in which they swam and pooped, and you have a complete cycle that leaves the ecosystem intact (figure 5.5).

When you scale up this system, it gets mind-boggling. In 2015 I spent quite a bit of time hanging around aquaculture conferences. At the World Aquaculture Congress held in Adelaide, I wandered through many sessions,

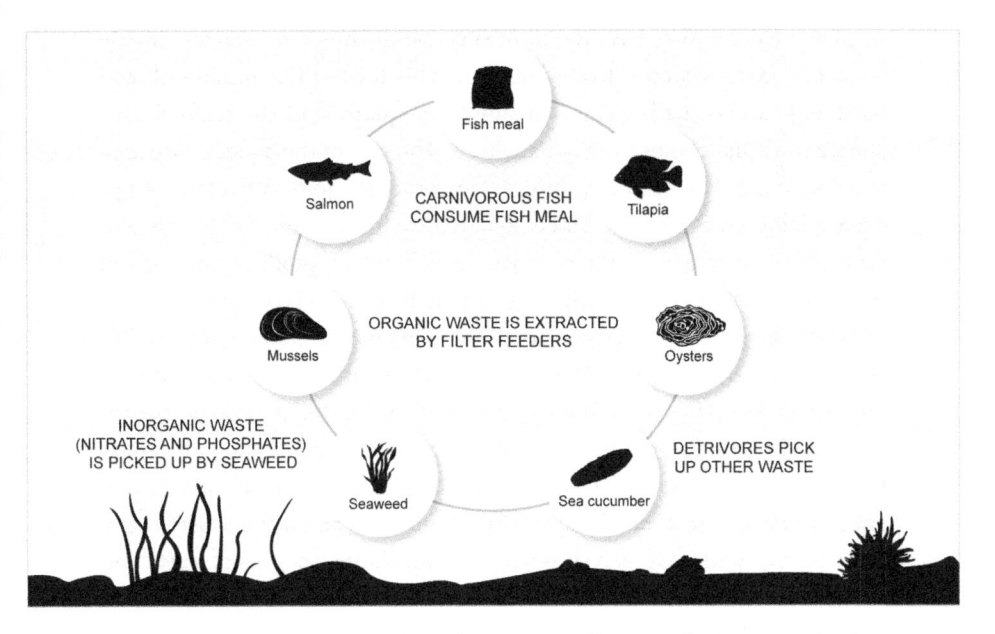

Fig. 5.5. Integrated multitrophic aquaculture system. Illustration by Morgan Richards.

gawked at the snazzy technology on display, and began to understand some of the excitement of aquaculture scientists. The buzzword is IMTA, or the integrated multitrophic aquaculture model that Chopin has done much to champion. He explains the concept: "Multi-trophic refers to the incorporation of species from different trophic or nutritional levels in the same system" (Barrington, Chopin, and Robinson 2009). In more elaborated terms, "IMTA is a practice in which the by-products (wastes) from one species are recycled to become inputs (fertilizers, food and energy) for another. Fed aquaculture species (e.g. finfish/shrimps) are combined, in the appropriate proportions, with organic extractive aquaculture species (e.g. suspension feeders/deposit feeders/herbivorous fish) and inorganic extractive aquaculture species (e.g. seaweeds) for a balanced ecosystem management approach" (Barrington, Chopin, and Robinson 2009, 7).

At the congress, there are sessions upon sessions about Chinese experiments in IMTA. In this area, the Chinese are huge in every way—their scientists are regarded as pioneers and their aquaculture production is enormous. In 2010, China produced 65 percent of the total aquaculture output in the world, 80 percent of the shellfish production, and 59 percent of the seaweed. While it is easy to stigmatize China as the biggest consumer of fish meal and fish oil, they have the most successfully implemented and highly advanced IMTA systems. In the long run, this lessens the amount of feed used and extends its benefits. Already three-quarters of the seafood consumed in China comes from aquaculture. The size of the aquaculture area is astonishing. For instance, Sungo Bay contains 130,000 hectares—130 square kilometers—of combined aquaculture production. In this system, typically the feed goes to the carnivorous fish, which produce organic and inorganic waste. Extractive filter feeders such as mussels or oysters eat the organic waste, and seaweed consumes the inorganic (nitrates and phosphorus). At the bottom a detritivore—an eater of detritus—such as the sea cucumber picks up the rest. Other species such as lobster are also detrivores and could potentially be brought into the IMTA system. Fang Jianguang's team, pioneers in IMTA in China, warn that ironically the best practice in aquaculture is threatened by agricultural practices and urban waste (Fang et al. 2015). While this neat system can handle large amounts of waste and even be used to bioremediate the water, it cannot hold up against the pesticide runoff from intensive land-based agriculture.

Thierry Chopin hopes that this really could be a revolution in food production. He, like many, knows all too well that the green revolution in agriculture brought problems: "The green revolution has increased yields and increased productivity . . . but in the short term. Now soils are eroding and getting exhausted" (Ogden 2013, 699). The fallout of the green revolution in agriculture in developing countries is precisely what aquaculture needs to avoid. The use of monocrops went against the culture of agriculture and produced devastating results. In terms of aquaculture, IMTA attempts to learn more closely from nature: "Cultivating more than one trophic level such that the wastes from fed organisms such as fish are recaptured and converted to fertilizer, food and a source of energy for other crops . . . would mimic aspects of the more complex marine communities seen in nature" (Ogden 2013, 700).

The Age of Algae

Minuscule phytoplanktons have floated throughout this chapter.[3] One of the spin-offs of the new research and practice of integrated aquaculture is a renewed respect for seaweed, which is one of the oldest food crops on the planet. Fourteen-thousand-year-old middens in Chile demonstrate that it was widely eaten as a foodstuff (Yeoman 2014).

It is the stuff of life. These marine microorganisms are the cornerstone of several ecosystems—the ocean, of course, as well as the terrestrial ecosystem that depends on the health of the sea. If we want to be human centered, plankton provides the oxygen that we breathe. Seaweed may also provide one of the answers to reducing the impact of aquaculture by stripping out nitrogen, phosphorus, and carbon, and reoxygenating the water.

As I was to find out on a trip down the south coast to visit a leading phycologist, Pia Winberg, seaweed may provide a way of making aquaculture sustainable. What if seaweed were the food that farmed fish ate? I first met Pia at the Fifth Congress of the International Society for Applied Phycology, which she organized in Sydney. Pia had organized a seaweed spread, which started with cocktails and seaweed canapés. They were very popular among this crowd of phycologists. I managed to snag one of the last oysters with ulva. I missed out on the Algae Mary, which incorporates oysters, chlorella powder, a Chinese white spirit, lemon juice, salt, and pepper. The cookbook Pia's research inspired was launched at the reception. *Coastal Chef: The Culinary Art of Seaweed and Algae in the 21st Century* is a collaboration with chefs working up and down Australia's New South Wales coast whom Pia persuaded to make dishes out of seaweed (Tinellis 2014). The recipe for the oyster shots describes them as a "balance [of] the briny sweetness of the algae-fed oysters with the heat of the spirits and the mild savory palate of the Chlorella powder" (Tinellis 2014, 66). Others include the gorgeous-looking Deep Sea Sour that has Amaretto, lemon juice, egg white, and blue phycocyanin spirulina extract, making it the most divine cobalt blue.

Pia's background is in marine systems ecology—not mixology—and her research focuses on marine food production systems sustainably integrated with the coastal and marine environment. She has left purely academic life to launch her own company, Venus Shell Systems, with the aim of making IMTA into a viable commercial enterprise. She works closely with a Sydney chef, Jared Ingersoll, on the Phyco Food Co., an online venture to sell and promote seaweed. Pia works endlessly doing the science, enticing investors, and promoting the benefits of algae. She has been featured in glowing terms in the business pages of the *Australian* newspaper: "If there are ever farms of waving fronds of seaweeds around the Australian coast, it will be mostly due to the persistence and research of that erudite mermaid, Dr. Pia Winberg" (Newton 2014).

Seaweed can be used as an ingredient in aquaculture feeds because it is high in protein and has omega-3 fatty acids as well as bioactive carbohydrates. The feed conversion rates for farmed carnivorous fish are improving. It is now estimated that it takes 6.8 kilograms of feed to produce 1 kilo of beef, whereas for fish it can be as low as 1.2 kilos for 1 kilo (Bourne 2014). Other modeling takes into account the carbon footprint, and of course beef is much higher. The problem is that for cost reasons there is a steady shift toward using soy products in fish feed. This takes us back to the same problem we encountered when the pigs were prevented from eating pig swill. Not only does it outsource the problem to places like Argentina, it does strange things to fish. Feeding soy to a salmon denatures it, whereas basing the feed on seaweed "gives it back its marine profile." Quite simply, it turns it back into a fish. As Pia argues, there are widespread human-fish benefits.

The global trend is to use soybean meal in aquaculture feeds, but there are other ways to incorporate protein and nutrition that keep the marine food chain intact—Indonesia and the Philippines developed seaweed industries as recently as the 1980s. Now there are dedicated seaweed villages where the income provides the resources for educating children. In some regions, seaweed has become vital for hundreds of thousands of families.

Pia's goal is to "build a mini-ecosystem by matching species that use each other's waste streams in which seaweed is one of the most important components" (ABC 2014). When I was down at Pia's lab, she showed me the prototype. Next to her lab in the Shoalhaven is Manildra, one of the world's largest producers of wheat starch. This produces a lot of carbon, which Winberg has harnessed in her seaweed production. Standing in the backyard outside her lab, I feel like I am witnessing a miracle. A large tub has CO_2 pumped in one end, and as Pia runs a sieve through the bubbling waters, there it is—bright green algae—new life.

Fish Relatedness

It is a fact of life that things are connected. What is remarkable is where, why, and with what effects human-fish-related connections are being made and remade. Sardines tell of the rise and fall of industry along the California coast, and menhaden are a stressed reminder of the American fish meal industries on the East Coast. It could be told as a narrative arc tracing the demise of a keystone species rendered fodder for fish and poultry farming, which culminates in the destruction of ecosystems. Telling the story this way emphasizes the inevitability of death and demise. Sometimes this is the case—the little fish upon which predators rest are at risk, imperiled by the appetites of some humans. But this narrative framing would miss "the complexities, frictions, intractabilities and conundrums of 'matter in relation'" (Abrahamsson et al. 2015, 9–10). The case of the Peruvian anchoveta affords a very different picture. Here we see, as Abrahamsson et al. argue, that "'doing' is a distributed achievement" (2015, 10). Dedicated tinkering with laws and regulations means that this little fish may be rejoined into a wider fish-related metabolic intimacy that helps to feed the poor and not just the wealthy omega-3 eaters of capsules. Certainly there is human agency here. A team composed of a cook, a biologist, and a designer helps to catalyze a different distribution of fish-related matters. Now some of those little fish do not end up pulped as meal and oil, matter fit only to feed to animals and other fish. They are instead "(re)valued as foodstuffs . . . as (re)formed food" (Coles and Hallett 2013, 157).

Of course, the reality remains that, as Brunner et al. argue, there will continue to be "conflicts between growing human demand for fish, and the need for sustainable fisheries that protect marine ecosystems and promote social and environmental justice" (2009, 93). By 2030, fish farms will provide two-thirds of the global fish supply. As Abrahamsson et al. (2015) state, "As human eaters organize themselves in complex sociomaterial ways, the fish they eat has become entangled with long distance trade routes." We have seen ample evidence of this as these fish-human relations move, and will do so increasingly, across North-South boundaries. As a crucial form of income for the Global South, little fish may be part of remaking that very geopolitical distinction. Or if entities such as the European Union continue to allow supertrawlers to pillage the fisheries off the west coast of Africa and elsewhere, the global gulf in access to income and decent nutrition will widen. As Mansfield (2010) puts it, the material production of fish could produce these and other recursive relationships, which will then occasion others. In the turn to aquaculture, we should remember Columella's point made millennia ago: that we should make "the fish feel their captivity as little as possible." In their multitrophic domain of scallops, mussels, sea cucumber, and algae, they are conjoined in a more fish-centered metabolic intimacy. This is a mode of doing that takes from the past and could remodel the very way in which we consider our bodily relatedness in the lives of little (and big) fishes.

These narratives are in the making. And it matters how we tell these stories. The little fish in this chapter are all recursively involved in multiple tales of human-fish relations. They—we—relate and interrelate in metabolic intimacy. To go back to Erdrich's fish that missed the buffalo for its ticks and its dung, the fish in this chapter effortlessly cross and reshape the boundaries that can make a difference between eating the ocean and eating it well.

Conclusion REELING IT IN

I love the Sydney Fish Market although it is not particularly lovable. It's a bit smelly, and more than a little scruffy. Pelicans and ibis have joined the throngs of marauding seagulls as they pick through the garbage and throw fries around (figure C.1). It sits between two concrete manufacturing factories overlooking Blackwattle Bay—which looks pretty murky at the moment. But it bustles with fish and people. It is the largest fish market in the southern hemisphere and one of the biggest in the world, although you wouldn't think so to look at it. At 4:30 in the morning, buyers start inspecting the wares at the wholesale side of the market. The auction starts at 5:30 AM. There are six hundred registered buyers, who buy sixty-five tons of fish and seafood each day, except Christmas Day, the one and only day of the year that the market closes. Some of the fish will go to the top-end restaurants, and some to small fishmongers in suburban shopping centers. Most of it goes to the large family-owned outfits like de Costi's and Claudio's.

During normal hours, the market brings together ordinary shoppers like me and masses of tourists. Buses disgorge tourists mainly from China. The Sydney Fish Market features highly on mainland Chinese organized group tours of Sydney. As you walk into the market, which is like one long corridor with shops and stalls on either side, it can at times feel like you're part of a king tide of bodies all searching for the best fish. People often don't get farther than the first shop. Large fish tanks: the crustaceans try to hide behind each other. Abalones the size of soup plates are plastered against the tanks, trying to escape. You choose what you want, and with some fanfare

Fig. C.1. Pelicans at Sydney Fish Market. Photograph by author.

the massive, and very elderly, lobster or mud crab is taken out, displayed for your satisfaction and that of your friends. Many selfies are taken, and then it's cooked however you want it. Families and groups of friends eat on rickety tables and uncomfortable plastic seats next to the queue for sashimi, and around the corner from the public toilets.

Most of the fish workers are Chinese and speak Cantonese, Mandarin, and English. At the entrance, a South Sea Islander woman dry-shucks oysters so you don't lose the juices. It's harder work than the practice of shucking under running water, but they are the best. She calls me "honey," and I give her a big smile. She shucks up to two thousand oysters a day and probably smiles at each oyster and its future consumer. Beyond her is the billboard dedicated to the local fishing fleet. They are all Italian—one big family of Bagnatos who started leaving the fishing village of Bagnara in Calabria before World War II. The few non-Bagnato names are the ones who married into the family.

There's something about the Sydney Fish Market that gives a unique sense of Sydney—an Asian-facing city that melds the earlier waves of Italian, Greek, and Vietnamese migrants with the smell of fish. Leading up to Christmas, you see more Anglo families coming to buy that quintessential Australian festive meal—prawns destined for the barbie. Every year I get on my bike and join the thousands in the fish-buying marathon (when the market stays open for thirty-eight hours). They sell over 170,000 tons of prawns. The queues are often fifteen deep, but there's camaraderie among seafood shoppers that you won't get in the department stores in town.

Some three hundred different species of fish and crustaceans are sold. People are rarely indifferent to what they buy. These are fish from Australia and New Zealand, with a very small percentage of imported fish from Thailand. Value-added fish—crumbed or batter-dipped prawns or those horrible rings of calamari that taste like rubber bands—are usually imported because of labor costs. By law, fish have to be labeled by country of origin when they are sold fresh but not when they are served cooked. They don't have to say which part of Australia the seafood comes from, but you get to know that the freshwater crayfish will be from Western Australia, and the Pacific oysters will be from Coffin Bay in South Australia, and the Sydney rocks from either north of the Hawkesbury River or down the south coast. But I have no idea of where exactly these stern-looking red snapper are from (figure C.2).

Is it sustainable? Well, yes and no. The future of Sydney Fish Market itself is seemingly up in the air. With skyrocketing prices for land with water views, developers wait backstage hoping that the government will loosen the regulations. The market is a strange amalgamation. It is a working port, a wholesale fish market, and a retail market. The land is publicly owned, but the two shareholders, the Catchers Trust and the Sydney Fish Market Tenants and Merchants Pty Ltd, have dominated the decisions. They are averse to change, and many are also worried that if the market were developed they would lose the real fish business of the wholesale market. The area is now slated for major change, with rumors of thirty-two-story apartment buildings on the edge of the market. But nothing much seems to be happening. The Chinese fishing and real estate conglomerate Dahua already owns a 25 percent stake in Sydney Fish Market. The sustainability and the very existence of the market are up for grabs, it seems.

Are the fish sustainable? The website for the Sydney Fish Market (2015) states, "Up to one hundred sustainable species traded every day." While that

Fig. c.2. Red snapper at Sydney Fish Market. Photograph by author.

sounds pretty good, there is no indication of what they mean by "sustainable." The Marine Stewardship Council lists only six certified Australian fisheries, which includes two prawn fisheries and one rock lobster fishery. The remaining are toothfish fisheries located in the Australian Subantarctic off Heard and McDonald Islands. The fisheries may be deemed sustainable, but their location is over 2,500 miles southwest from Perth. Getting to the fishing grounds takes over two weeks, and the amount of fuel used would hardly be deemed sustainable in a wider sense.

As I've argued across this book, what constitutes sustainability is a fraught and complex set of issues. It's even trickier when sustainability has to cover the entire relatedness of fish, humans, and ocean, not just biological or economic relations. I've argued that some species should not be eaten, whereas it is important to eat others such as sardines and anchovies, the little fish that feed the big fish. And in general farmed oysters and other bivalves are a good way to get protein without overly damaging the ocean. But I have also tried to get at something more than just what is or what isn't good to eat. *Eating the Ocean* proceeds with a sensibility that comes from acknowledging the awesome nature of the ocean, and that respects its inhabitants and dependants. What that entails varies, but it can never be in isolation. An ethics of the oceanic refuses the morality of piecemeal denunciation, of a food politics impoverished through simplification. Here in the space of an aging and not very pretty fish market you see glimpses of fish and human relatedness. People bend over the fish, marvel at their colors, want to know where they are from and how to cook them. However fleeting, there is an attention to and a respect for these fish as storied and cultured entities. Equally, there is a feeling of respect for the fishermen. People walk out to the boats and look on as the fishers wash down the boats and sort their nets. They take selfies against the background of the billboard that features photographs of the local fishing fleet.

What I hope to do in this book is to bring a new sensibility to questions about human relations with nonhumans, another take on what the oceanic more-than-human might afford. I'm not sure if I propose a new ontology, although I certainly argue for a different way of approaching how we think about eating the ocean. From the outset, I attempt to relate the wonder of the ocean as a first principle in how to eat the ocean better. Against Coleridge's (1834) benighted sailor, we are never "alone, alone on a wide, wide sea." The more-than-human, if it is to be meaningful as a perspective, makes us confront again and again the relatedness of all entities. And while some may say that the best way of honoring that relatedness is not to eat fish, as I've argued, this is not a solution. I won't force-feed people sardines or anchovies or oysters—what a waste that would be. But keep in mind the pressures of a growing population; be aware of the state of land-based agriculture; be informed of the advances in sustainable systems such as integrated marine trophic aquaculture; be mindful of the millions who work with the sea. This is why this book focuses on noticing detail, relating stories, histories, environments, and tastes. Try to eat the ocean better. Try to eat with the ocean.

NOTES

Introduction
1   This program was part of the Hawke Research Institute at the University of South Australia, and I thank them for the funding. I learned much from colleagues working in the area of rural studies there, especially Jen Cleary, Lia Bryant, and Guy Robinson. My thanks to Lisa Slater for her help and company on research travel around South Australia. Some of the areas we studied included the politics of food production and consumption; regionality within the global; food security; Indigenous food enterprises; terroir, including water, soil, climate; and new markets, old problems.
2   Thank you, Janet Dibb-Leigh.
3   Australian Research Council Discovery Project, "Sustainable Fish: A Material Analysis of Cultures of Consumption and Production," DP140101537.
4   I have learned much from my student Kate Johnston's doctoral work on tradition and culture in Italian tuna fisheries.
5   FAO (2015b); see also Watson (2015).
6   Because theirs is very much a team project, when I cite publications from this project I use the term "Mol and her team" as well as the name of the person who happens to be the first author on any individual article.

1. An Oceanic Habitus
1   See my (Probyn 2012) argument about eating "roo" as well as Emma Roe's (2006) article about "things that become edible" for other takes on this. In Nicholas Röhl and Greta Scacchi's campaign Fishlove, this is precisely what famous and used-to-be famous celebrities do—cuddle fish (see http://www.fishlove.co.uk/).
2   It goes without saying that choice is the byword of neoliberalism. While I am wary of how neoliberalism yokes together issues of very different orders with an overwhelming emphasis on the actions of the individual—the power of the fork—I

have long had an interest in the deployments of ideas about choice. Years ago I wrote about "choiceosie," a neologism that encompasses the social class and cultural capital that empowers people to think that they can freely choose and denies the materiality that one has to choose something over something else (Probyn 1995).

3   The individual transferable quota system allocates shares to individuals on the basis of what the total allowable catch is that year for certain species. As I discuss across the book, it is central to the management of certain fish species, especially high-value ones like lobster or bluefin. It is, however, far from uncontested or perfect. Many see the system as a market device for regulating fisheries that privatizes the commons of inshore and deep-sea fishing.

4   In 2014 I interviewed Fiona Wheatley, the manager of Marks and Spencer's Plan A (for the sustainability of natural resources). It was an eye-opening interview. I am among the legions that love M&S; however, I now have a firmer basis for that affection. They really seem to have figured out how to go about rendering the entire fish food chain on the way to complete transparency. They have one huge advantage over other retailers: Their food products are almost 100 percent home brand, and therefore they are in a powerful position to insist on sustainable seafood throughout the entire process, including value-added products such as fish cakes. In the United Kingdom, their Forever Fish is well known to the public (every time you forget your own shopping bag and need to buy a plastic bag, you pay a small amount to the cause). The campaign has three objectives: to "help to protect and save our precious sea life, oceans and beaches for future generations to enjoy," to "encourage eating of lesser known and British fish species, without compromising on quality," and "involving volunteers in cleaning our beaches, and teaching their children about fish." I was particularly struck by their desire to help fisheries change. Rather than simply going by the stop sign system and avoiding less sustainable stocks, they actively work with those who have a red rating but who are working on improving. According to the manager, it is more important to work with fisheries to become sustainable—"Don't ban stuff; make it better" (personal communication, April 16, 2014).

5   The question of caring within the more-than-human realm complicates the status as well as the relationship between the object and subject of care. As María Puig de la Bellacasa (2010, 152) notes, we also need to think about "what sort" of care. In her project "The Eating Body," Annemarie Mol brings her previous work on care within health practices (Mol 2008) into the realm of eating (Yates-Doerr 2012). Mara Miele has done much both at a conceptual level and in policy to foster a wider sense of what caring for what we eat would mean. Through her involvement in an EU study for better guidelines about the welfare of farmed animals, Miele (2011) has written about how to understand "the happiness of chickens," and with her colleague Adrian Evans contemplates the wide dimensions of an "ethics and responsibility in care-*full* practices of consumption" (Miele and Evans 2010).

6   In her review of Raymond Gaita's *The Philosopher's Dog*, Val Plumwood takes issue with his assertion "I cannot, and I know of no one else who can feel the same about animal killing as about human killing" (Gaita 2002, 211). Plumwood (2007)

replies, "I have to say that I personally *feel* the same outrage at the mass murder and machine gunning of seal colonies and dolphin pods as I do about similar mass killings of humans." I have to say that such arguments leave me conflicted. Might there not be a middle ground?

2. Following Oysters, Relating Taste

1. Throughout this chapter, I refer mainly to *Crassostrea* within the Ostreidae family—true oysters. Pacific oysters, *Crassostrea gigas*, are the preferred species for farming because they grow much more quickly than rocks, or *Saccostrea*. Pearl oysters, *Pteriidae*, are not considered true oysters. In Broome in northern Western Australia, where pearling is an important business, you can find pearl meat for sale to eat, which is tasty but very tough.
2. As my student Helen Greenwood so eloquently describes in her thesis on women food writers.
3. The verse is thought to have been written by P. G. Wodehouse (see Madame Eulalie's Rare Plums 2013).
4. The new colony was built with oyster shells. The making of "shelly mortar" consumed vast amounts of shells, sadly often taken from "the great middens—often massive piles of discarded shells, bones and other detritus—built up over thousands of years by local Aboriginal groups" (Sydney Living Museums 2012).
5. It is all the more astonishing that Fisher wrote this at the age of sixteen. Much of her writing was somewhat out of kilter with her life. Her extraordinary book *Consider the Oyster* was conducted against the backdrop of her second marriage and most intense love with Dillwyn Parrish, who committed suicide in 1941.
6. Thanks to my late brother, Stephen Probyn, for this description of the British Empire, which was indeed colored pink on world maps.
7. This appeared in the "Philosophy" section on the company website when the company was employee owned. It now has a slightly different text that refers only to the welfare of the oysters and their marine habitat, and not to that of the people who tend them (see Loch Fyne Oysters 2015).
8. For more information on the dealings of Greene King, see Greene King (2014).

3. Swimming with Tuna

1. Kate Barclay and Charlotte Epstein note that "the eating of four-legged animals was banned by the government from 675 CE, largely for religious reasons. Although the ban was lifted in the 1860s, cultural patterns and the economics of raising animals for food meant consumption of animals other than chickens or fish remained low for another century" (2013, 220).
2. The interviews were conducted in Port Lincoln in February 2010 and March 2011. The interviews in Ulladulla were in April 2011. Unless otherwise noted, I have with their permission used real names. My thanks to Len Stephens, the CEO of the Collaborative Research Centre on Seafood, for his help in contacting people in the industries. And of course my thanks to those I interviewed. The interviews were open ended and lasted for one to two hours.

3   For a focused analysis of Japan's tuna industry and its links to development strategies in the Pacific, see Barclay (2008). As Barclay and others have pointed out, fish and fishing are particularly important to Japan given its very limited capacity to feed itself from terrestrial sources. Historically this has produced very close relations between the government and Japanese fisheries (Barclay and Koh 2006). For another view of the special meaning that fish has in Japan, see also Theodore Bestor's ethnographies of sushi in Japanese culture as well as his account of Tsukiji Fish Market (Bestor 2000, 2004).
4   The International Commission for the Conservation of Atlantic Tunas controls Atlantic bluefin.

4. Mermaids, Fishwives, and Herring Quines

1   Thanks to Jennifer Biddle for bringing this to my attention.
2   In Whatmore's listing of what the more-than-human entails or compels, we have a fourfold template. This includes the by-now accepted refrain about practice, not discourse. Other elements are affect, not meaning, and a "shift from a focus on the *politics of identity* to *the politics of knowledge*" (Whatmore 2006, 603–4).
3   Haraway's remarks about touch also can be seen in the longer context of feminist thinking and research on embodiment and more recently on affect, which is again vast, but the genealogies of which should not be forgotten. While I have written in more detail about the ethics of embodied ethnographic research that my longer project uses (Probyn 2015), I am always mindful of Angela McRobbie's (1982) early warning not to take for granted the access that women researchers have to other women in her "Politics of Feminist Research."

5. Little Fish

1   The people behind this website describe themselves as follows: "The George Mateljan Foundation, a not-for-profit foundation with no commercial interests or advertising, is a new force for change to help make a healthier you and a healthier world" (George Mateljan Foundation 2015). The DRI that this site uses comes from the U.S. Department of Agriculture.
2   Jodi Frawley was the chief organizer. See Maricultures Environmental Research, accessed March 2, 2015, http://sydney.edu.au/environment-institute/mer/.
3   For more on the phrase "the age of algae," see Chopin (2015).

REFERENCES

ABC. 2013. "Three Men and a Boat." *Landline*, July 4. http://www.abc.net.au/landline/content/2013/s3731527.htm.
ABC. 2014. "Slimy Harvest." *Landline*, October 5. http://www.abc.net.au/landline/content/2014/s4002127.htm.
Abrahamsson, Sebastian, Filippo Bertoni, Annemarie Mol, and Rebeca Ibáñez Martín. 2015. "Living within Omega-3: New Materialism and Enduring Concerns." *Environment and Planning D: Society and Space* 33 (1): 4–19.
Ads of the World. 2014a. "Sea Shepherd Conservation Society: Tuna, 2." Accessed October 9. http://adsoftheworld.com/media/print/sea_shepherd_conservation_society_tuna.
Ads of the World. 2014b. "WWF Bluefin Tuna Overfishing: Rhino." Accessed October 9. http://adsoftheworld.com/media/print/wwf_bluefin_tuna_overfishing_rhino.
Aguilar, Lorena, and Itza Castañeda. 2001. *About Fishermen, Fisherwomen, Oceans and Tides: A Gender Perspective in Marine-Coastal Zones*. San José: IUCN.
Alaimo, Stacy. 2012a. "States of Suspension: Trans-corporeality at Sea." *Interdisciplinary Studies in Literature and the Environment* 19 (3): 476–93.
Alaimo, Stacy. 2012b. "Sustainable This, Sustainable That: New Materialisms, Posthumanism and Unknown Futures." *PMLA* 127 (3): 558–64.
Allen, Robin. 2010. *International Management of Tuna Fisheries: Arrangements, Challenges and a Way Forward*. Rome: Food and Agriculture Organization of the United Nations.
Allewaert, M., and Michael Ziser. 2012. "Preface: Under Water." *American Literature* 84 (2): 233–41.
Andersen, Hans Christian. 2006. *Stories from Hans Andersen*. Project Gutenberg. http://www.gutenberg.org/files/17860/17860-h/17860-h.htm.
Anderson, Ben, and John Wylie. 2009. "On Geography and Materiality." *Environment and Planning A* 41 (2): 318–35.

Arora-Jonsson, Seema. 2011. "Virtue and Vulnerability: Discourses on Women, Gender and Climate Change." *Global Environmental Change* 21 (2): 744–51.

Authority Nutrition. 2015. "About Authority Nutrition." Accessed January 12. http://authoritynutrition.com/about/.

Balint, Ruth. 2012. "Aboriginal Women and Asian Men: A Maritime History of Color in White Australia." *Signs* 37 (3): 544–54.

Banse, Karl. 1990. "Mermaids—Their Biology, Culture, and Demise." *Limnology and Oceanography* 35 (1): 148–53.

Barad, Karen. 2012. "Nature's Queer Performativity." *Kvinder, Køn og Forskning* (1–2).

Barclay, Eliza. 2009. "Chef Gaston Acurio Carves Peruvian-Flavored Empire." *Los Angeles Times*, January 21.

Barclay, Kate. 2008. *A Japanese Joint Venture in the Pacific: Foreign Bodies in Tinned Tuna*. London: Routledge.

Barclay, Kate, and Charlotte Epstein. 2013. "Securing Fish for the Nation: Food Security and Governmentality in Japan." *Asian Studies Review* 37 (2): 215–33.

Barclay, Kate, and Koh Sunhui. 2006. "Marketization of Japanese Governance? A Case Study of Long Line Tuna Fisheries." In *Regionalization, Marketization and Political Change in the Pacific Rim*, edited by James Goodman, 347–83. Guadalajara, Mexico: Editorial Centro Universitario de Sociales y Humanidades, University of Guadalajara Press.

Barrington, Kelly, Thierry Chopin, and Shawn Robinson. 2009. "Integrated Multitrophic Aquaculture (IMTA) in Marine Temperate Waters." In *FAO Fisheries and Aquaculture Technical Paper, No. 529*, edited by D. Soto, 7–46. Rome: FAO.

Barron, James. 2008. "Warnings Don't Deter Lovers of Sushi." *New York Times*, January 24.

Barthes, Roland. 1972. *Mythologies*. Translated by Annette Lavers. New York: Hill and Wang.

Bavington, Dean. 2008. "Managing to Endanger: Creating Manageable Cod Fisheries in Newfoundland and Labrador, Canada." *MAST* 7 (2): 99–121.

Bear, Christopher, and Sally Eden. 2011. "Thinking Like a Fish? Engaging with Nonhuman Difference through Recreational Angling." *Environment and Planning D: Society and Space* 29 (2): 336–52.

Bennett, Jane. 2009. *Vibrant Matter: A Political Ecology of Things*. Durham, NC: Duke University Press.

Berkeley Wellness. 2014. "Omega-3 Supplements in Question." March 9. http://www.berkeleywellness.com/supplements/other-supplements/article/omega-3-supplements-question.

Berlant, Lauren. 2011. *Cruel Optimism*. Durham, NC: Duke University Press.

Bestor, Theodore C. 2000. "How Sushi Went Global." *Foreign Policy*, November–December.

Bestor, Theodore C. 2001. "Supply-Side Sushi: Commodity, Market, and the Global City." *American Anthropologist* 102 (1): 76–95.

Bestor, Theodore C. 2004. *Tsukiji: The Fish Market at the Center of the World*. Berkeley: University of California Press.

Birke, Lynda. 2002. "Intimate Familiarities? Feminism and Human-Animal Studies." *Society and Animals* 10 (4): 429–36.

BlueVoice. 2008. "A Shared Fate" [film]. http://www.bluevoice.org/webfilms_sharedfate.php.
Blyth, Matt, and Alaneo Gloor. 2013. *Drawing the Line* [film]. Sydney: Millstream.
Bourdieu, Pierre. 1984. *Distinction: A Social Critique of the Judgment of Taste*. Cambridge, MA: Harvard University Press.
Bourdieu, Pierre. 1990. *The Logic of Practice*. Stanford, CA: Stanford University Press.
Bourne, Joel K. 2014. "How to Farm a Better Fish." *National Geographic*. http://www.nationalgeographic.com/foodfeatures/aquaculture/.
Brown, Lester R. 2006. "Feeding Seven Billion Well." In *Plan B 2.0: Rescuing a Planet under Stress and a Civilization in Trouble*. New York: W. W. Norton.
Brunner, Eric J., Peter J. S. Jones, Sharon Friel, and Mel Bartley. 2009. "Fish, Human Health and Marine Ecosystem Health: Policies in Collision." *International Journal of Epidemiology* 38 (1): 93–100.
Burton, Valerie. 2012. "Fish/Wives: An Introduction." *Signs* 37 (3): 527–36.
Butler, Judith. 1990. *Gender Trouble*. New York: Routledge.
Callon, Michel. 1999. "Actor-Network Theory—the Market Test." In *Actor Network Theory and After*, edited by John Law and John Hassard, 181–95. Oxford: Blackwell.
Campling, Liam, and Elizabeth Havice. 2014. "The Problem of Property in Industrial Fisheries." *Journal of Peasant Studies* 41 (5): 707–27.
Cardozo, Karen, and Banu Subramaniam. 2013. "Assembling Asian/American Naturecultures: Orientalism and Invited Invasions." *Journal of Asian American Studies* 16 (1): 1–23.
Carroll, Lewis. 1872. "The Walrus and the Carpenter." Available at Jabberwocky, http://www.jabberwocky.com/carroll/walrus.html.
Channel 4 Britdoc Foundation. 2009. *"The End of the Line*: A Social Impact Evaluation." http://britdoc.org/uploads/media_items/theendoftheline-evaluationdocument.original.pdf.
Chesapeake Bay Foundation. 2014. "Great Shellfish of the Bay." Accessed August 26. http://www.cbf.org/about-the-bay/more-than-just-the-bay/creatures-of-the-chesapeake/eastern-oyster.
Chiang, Connie Y. 2008. *Shaping the Shoreline: Fisheries and Tourism on the Monterey Coast*. Seattle: University of Washington Press.
Chopin, Thierry. 2015. "The Age of Algae" [interview]. *Atlantic Business Magazine*, January 19. http://www.atlanticbusinessmagazine.net/the-age-of-algae/.
Christensen, Villy, Santiago de la Puente, Juan Carlos Sueiro, Jeroen Steenbeek, and Patricia Majluf. 2014. "Valuing Seafood: The Peruvian Fisheries Sector." *Marine Policy* 44: 302–11.
Coleridge, Samuel Taylor. 1834. "The Rime of the Ancient Mariner." Available at Poetry Foundation, http://www.poetryfoundation.org/poem/173253.
Coles, Benjamin, and Lucius Hallett. 2013. "Eating from the Bin: Salmon Heads, Waste and the Geographies That Make Them." *Sociological Review Monographs* 60: 156–73.
Connery, Christopher L. 1996. "The Oceanic Feeling and Regional Imaginary." In *Global/Local: Cultural Production and the Transnational Imaginary*, edited by Rob Wilson and Wimal Dissanayake, 284–311. Durham, NC: Duke University Press.

Cook, Ian, et al. 2004. "Follow the Thing: Papaya." *Antipode* 36 (4): 642–64.

Coutts, Karen Harvey, Alejando Chu, and John Krigbaum. 2011. "Paleodiet in Late Preceramic Peru: Preliminary Isotopic Data from Bandurria." *Journal of Island and Coastal Archaeology* 6 (2): 196–210.

Cuthbert, Pamela. 2014. "Size Doesn't Matter." *Doctor's Review*, February. http://www.doctorsreview.com/features/size-doesnt-matter/.

Davis, Dona Lee. 1986. "Occupational Community and Fishermen's Wives in a Newfoundland Fishing Village." *Anthropological Quarterly* 59 (3): 129–42.

Davis, Dona Lee. 1993. "When Men Become 'Women': Gender Antagonism and the Changing Sexual Geography of Work in Newfoundland." *Sex Roles* 29 (7–8): 457–75.

Davis, Dona Lee, and Siri Gerrard. 2000. "Introduction: Gender and Resource Crisis in the North Atlantic Fisheries." *Women's Studies International Forum* 23 (3): 279–86.

Debelle, Penelope. 2006. "Fishy Behaviour Doesn't Worry the Millionaires of Port Lincoln." *The Age*, August 19.

de Lauretis, Teresa. 1987. *Technologies of Gender: Essays on Theory, Film, and Fiction*. Bloomington: Indiana University Press.

Deleuze, Gilles. 1992. "Ethology: Spinoza and Us." In *Incorporations*, edited by Jonathan Crary and Sanford Kwinter. New York: Zone.

Deleuze, Gilles, and Félix Guattari. 2004 [1980]. *A Thousand Plateaus*. Translated by Brian Massumi. London: Continuum.

Department of the Environment. 2014. "Commonwealth Marine Reserves—Allowed Activities." Accessed March 17. http://www.environment.gov.au/topics/marine/marine-reserves/overview/allowed-activities.

Duarte, Carlos M., Marianne Holmer, Yngvar Olsen, Doris Soto, Núria Marbá, Joana Guiu, Kenny Black, and Ioannis Karakassis. 2009. "Will the Oceans Help Feed Humanity?" *BioScience* 59 (11): 967–76.

Dunphy, Shay. 2013. "'Waterford Parted from the Sea': The Irish in Newfoundland." *The Irish Story*, February 17. http://www.theirishstory.com/2013/02/17/waterford-parted-from-the-sea-the-irish-in-newfoundland/#.VMXNSMZLHLU.

Dutta, Abhijit. 2015. "Tsukiji, Tokyo: Among the Believers." *Live Mint*, March 28.

Dwyer, June. 2005. "Yann Martel's *Life of Pi* and the Evolution of the Shipwreck Narrative." *Modern Language Studies* 35 (2): 9–21.

East Lothian Museums. 1999. "The Silver Darlings." http://www.eastlothianmuseums.org/exhibitions/harvest/sea2.htm.

Eat the Seasons. 2008. "Eat Oysters." Accessed October 29. http://www.eattheseasons.co.uk/Articles/oysters.php.

Ensor, Sarah. 2012. "Spinster Ecology: Rachel Carson, Sarah Orne Jewett, and Nonreproductive Futurity." *American Literature* 84 (2): 409–35.

Erdrich, Louise. 2012. *The Round House*. New York: Harper.

Everaert, Claudine, and Gertjan Zwanikken. 2010. *Sea the Truth* [film]. Amsterdam: Alalena.

Faier, Lieba, and Lisa Rofel. 2014. "Enthographies of Encounter." *Annual Review of Anthropology* 43: 363–77.

Fairfood. 2015. "Caught in a Trap." http://www.fairfood.org/wp-content/uploads/2015/04/Caught-in-a-trap.pdf.
Fang Jianguang, Zhang Jihong, Jiang Zengjie, and Mao Yuze. 2015. "Development of Integrated Multi-trophic Aquaculture in China." World Aquaculture Society. Accessed March 5. https://www.was.org/documents/MeetingPresentations/WA2014/WA2014_0592.pdf.
FAO. 2015a. "Fisheries." Accessed April 9. http://www.fao.org/fisheries/en/.
FAO. 2015b. "General Situation of World Fish Stocks." Accessed April 9. http://www.fao.org/newsroom/common/ecg/1000505/en/stocks.pdf.
Fisher, M. F. K. 1990. *The Art of Eating*. New York: Hungry Minds.
ForArgyll.com. 2012. "Loch Fyne Oysters Sold to Scottish Seafood Investments." http://forargyll.com/2012/02/loch-fyne-oysters-sold-to-scottish-seafood-investments/.
Fotsch, Paul. 2004. "Tourism's Uneven Impact: History on Cannery Row." *Annals of Tourism Research* 31 (4): 779–800.
Foucault, Michel. 1986. "Of Other Spaces." *Diacritics* 16: 22–27.
Foucault, Michel. 1998. "Polemics, Politics and Problematizations." In *Essential Works of Foucault*, vol. 1, *Ethics: Subjectivity and Truth*, edited by Paul Rabinow, 111–20. New York: New Press.
Frank, Peter. 1976. "Women's Work in the Yorkshire Inshore Fishing Industry." *Oral History* 4 (1): 57–72.
Franklin, H. Bruce. 2007. *The Most Important Fish in the Sea: Menhaden and America*. Washington, DC: Island.
Freon, Pierre, Marilu Bouchon, Gilles Domalain, Carlota Estrella, Federico Iriarte, Jérôme Lazard, Marc Legendre, Isabel Quispe, Tania Mendo, Yann Moreau, Jesus Nuñez, Juan Carlos Sueiro, Jorge Tam, Peter Tyedmers, and Sylvestre Voisin. 2010. "Impacts of the Peruvian Anchoveta Supply Chains: From Wild Fish in the Water to Protein on the Plate." GLOBEC *International Newsletter*, April, 27–31.
Freud, Sigmund. 2002. *Civilisation and Its Discontents*. London: Penguin.
Future of Fish. 2014. "Breakthrough Aquaculture." http://www.futureoffish.org/sites/default/files/docs/resources/Aquaculture_Report_FoF_2014.pdf.
Gadamer, Hans-Georg. 1997. *Truth and Method*. New York: Continuum.
Gaita, Raimond. 2002. *The Philosopher's Dog*. Melbourne: Text Publishing.
Gatens, Moira. 1996. "Sex, Gender, Sexuality: Can Ethologists Practice Genealogy?" *Southern Journal of Philosophy* 35 (s): 1–19.
Gender in Aquaculture and Fisheries. 2014. "The Long Journey to Equality: Report on the 5th Global Symposium on Gender in Aquaculture and Fisheries, 13–15 November 2014, Lucknow, India." http://genderaquafish.org/gaf5-2014-lucknow-india/.
George Mateljan Foundation. 2015. "Eggs, Pasture-Raised." World's Healthiest Foods. Accessed March 2. http://www.whfoods.com/genpage.php?tname=foodspice&dbid=92.
Gibson-Graham, J. K. 2011. "A Feminist Project of Belonging for the Anthropocene." *Gender, Place and Culture: A Journal of Feminist Geography* 18 (1): 1–21.

Gilroy, Paul. 1993. *The Black Atlantic*. Cambridge, MA: Harvard University Press.

Goodman, David. 1999. "Agro-Food Studies in the 'Age of Ecology': Nature, Corporeality, Bio-politics." *Sociologica Ruralis* 39 (1): 17–38.

Goodman, Michael K. 2008. "Towards Visceral Entanglements: Knowing and Growing Economic Geographies of Food." Environment, Politics and Development Working Paper Series, no. 5. Kings College, London.

Goodman, Mike. 2015. "Food Geographies I: Of Relational Foodscapes and the Busyness of Being More-Than-Food." *Progress in Human Geography*, February 16.

Greenberg, Paul. 2009. "A Fish Oil Story." *New York Times*, December 15.

Greenberg, Paul. 2010. *Four Fish: The Future of the Last Wild Food*. New York: Penguin.

Greenberg, Paul. 2012. "An Oyster in the Storm." *New York Times*, October 29.

Greene King. 2014. "Results and Presentations." Accessed September 2. http://www.greeneking.co.uk/index.asp?pageid=43.

Grout, James. 2015. "*Piscinae*: Roman Fishponds." In *Encyclopaedia Romana*, http://penelope.uchicago.edu/~grout/encyclopaedia_romana/wine/piscinae.html.

Guthman, Julie. 2007. "Can't Stomach It: How Michael Pollan et al. Made Me Want to Eat Cheetos." *Gastronomica* 7 (3): 75–79.

Halliday, Stephen. 1999. *The Great Stink of London: Sir Joseph Bazalgette and the Cleansing of the Victorian Metropolis*. Gloucestershire: Sutton.

Hamilton, Gordon. 2011. "Sardines Return by the Millions to B.C." *Vancouver Sun*, April 6.

Haraway, Donna J. 2008. *When Species Meet (Posthumanities)*. Minneapolis: University of Minnesota Press.

Harbers, Hans, Annemarie Mol, and Alice Stollmeijer. 2002. "Food Matters: Arguments for an Ethnography of Daily Care." *Theory, Culture and Society* 19 (5/6): 207–26.

Hardin, Garrett. 1968. "The Tragedy of the Commons." *Science* 162 (3859): 1243–48.

Hardy, Anne. 2003. "Exorcizing Molly Malone: Typhoid and Shellfish Consumption in Urban Britain 1860–1960." *History Workshop Journal* 55 (1): 72–90.

Harold T. Griffin Inc. 2015. "Have You Ever Seen a Fat Lazy Norwegian?" Accessed June 6. http://www.htgriffin.com/norwege.htm.

Hawke, Shé. 2013. "Aquamater: A Genealogy of Water." *Feminist Review* 103: 120–32.

Hayes-Conroy, Allison, and Jessica Hayes-Conroy. 2008. "Taking Back Taste: Feminism, Food and Visceral Politics." *Gender, Place and Culture: A Journal of Feminist Geography* 15 (5): 461–73.

Hayward, Eva. 2010. "Fingeryeyes: Impressions of Cup Corals." *Cultural Anthropology* 25 (4): 577–99.

Hayward, Eva. 2012. "Sensational Jellyfish: Aquarium Affects and the Matter of Immersion." *differences: A Journal of Feminist Cultural Studies* 23 (3): 161–96.

Heise, Ursula K. 2014. "Ursula K. Heise Reviews Timothy Morton's *Hyperobjects*." *Critical Inquiry*, June 4.

Helmreich, Stefan. 2009. *Alien Ocean: Anthropological Voyages in Microbial Seas*. Berkeley: University of California Press.

Helmreich, Stefan. 2011. "Nature/Culture/Seawater." *American Anthropologist* 133 (1): 132–44.

Hennion, Antoine. 2007. "Those Things That Hold Us Together: Taste and Sociology." *Cultural Sociology* 1 (1): 97–114.

Here We Are. 2008. "Welcome to Here We Are . . . " Accessed November 4. http://www.hereweare-uk.com/aboutus.

Hoar, Rebecca. 2000. "Coming Up Fast: The Man Who Made the Oyster His World." *Management Today*, August 1.

Hogan, C. 2012. "Commensalism." In *The Encyclopedia of Earth*. http://www.eoearth.org/view/article/171918/.

Howard, Penny McCall. 2015. "What Wrecks Reveal: Structural Violence in Ecological Systems." In *Environmental Change and the World's Futures: Ecologies, Ontologies, and Mythologies*, edited by Jonathan Paul Marshall and Linda H. Connor, 196–213. London: Routledge.

Howarth, Leigh M., Callum M. Roberts, Ruth H. Thurstan, and Bryce D. Stewart. 2014. "The Unintended Consequences of Simplifying the Sea: Making the Case for Complexity." *Fish and Fisheries* 15 (4): 690–711.

Hughes, Johnson Donald. 2001. *An Environmental History of the World: Humankind's Changing Role in the Community of Life*. New York: Routledge.

Huxley, T. H. 1882. "Inaugural Address." Available at Huxley File, http://alepho.clarku.edu/huxley/SM5/fish.html.

IFFO. 2011. "Responding to Our Critics." http://www.iffo.net/system/files/Responding%20to%20our%20Critics%20-%20Fishmeal%20and%20Fish%20oil%20Production%20_EN_%20Jan%2011_.pdf.

Ingold, Tim. 1991. "Evolutionary Models in the Social Sciences." *Cultural Dynamics* 4 (3): 239–50.

Issenberg, Sasha. 2007. *The Sushi Economy: Globalization and the Making of a Modern Delicacy*. New York: Gotham.

Jacobsen, Rowan. 2008. *A Geography of Oysters: The Connoisseur's Guide to Oyster Eating in North America*. New York: Bloomsbury.

"Japan Urges Industry-Wide Reduction of Tuna Effort." 2010. *World Fishing and Aquaculture*, December 15. http://www.worldfishing.net/news101/regional-focus/japan-urges-industry-wide-reduction-of-tuna-effort.

Krummer, Corby. 2007. "The Rise of the Sardine." *Atlantic*, July–August.

Kurth, Peter. 1999. "Tipping the Velvet." *Salon*. http://www.salon.com/1999/07/30/waters/.

Labarte, Maria Luisa del Rio. 2012. "Peru: Outgoing Vice Minister Reports on Problems in Fishing Industry." *Peru This Week*, May 7.

Latour, Bruno. 2004. "How to Talk about the Body? The Normative Dimensions of Science Studies." *Body and Society* 10 (2–3): 205–29.

Latour, Bruno. 2005. *Reassembling the Social: An Introduction to Actor-Network Theory*. Oxford: Oxford University Press.

Law, John, and Annemaire Mol. 2008. "Globalisation in Practice: On the Politics of Boiling Pigswill." *Geoforum* 39 (1): 133–43.

Lee, Michael Parrish. 2014. "Eating Things: Food, Animals and Other Life Forms in Lewis Carroll's Alice Books." *Nineteenth-Century Literature* 68 (4): 484–512.

Lewis, Edward A. 2000. "VI. History of the Fishery." University Corporation for Atmospheric Research. http://www.ucar.edu/communications/gcip/m12anchovy/m12pdfc6.pdf.

Loch Fyne Oysters. 2015. "Our Philosophy." http://www.lochfyne.com/about-us/our-philosophy.

Lorimer, Hayden. 2005. "Cultural Geography: The Busyness of Being 'More-Than-Representational.'" *Progress in Human Geography* 29 (1): 83–94.

Lukin Fisheries. 2010. "About Dinko." Accessed March 12. http://www.lukinfisheries.com.au/tuna.htm.

Madame Eulalie's Rare Plums. 2013. "AVENGED!" http://www.madameulalie.org/punch/Avenged.html.

Majluf, Patricia. 2013. "The Very Elusive Win-Win-Win (A Story of Greed, Overfishing, Perceptions, Luck, Peculiar Circumstances, and Hopefully a Happy Ending)." Paper presented at the Changing Coastlines Symposium, Sydney Environment Institute, Sydney, Australia, November 7–8.

Mann, Anna, Annemarie Mol, Priya Satalkar, Amalinda Savirani, Nasima Selim, Malini Sur, and Emily Yates-Doerr. 2011. "Mixing Methods, Tasting Fingers: Notes on an Ethnographic Experiment." *Journal of Ethnographic Theory* 1 (1): 221–43.

Mansfield, Becky. 2010. "Is Fish Health Food or Poison? Farmed Fish and the Material Production of Un/Healthy Nature." *Antipode* 43 (2): 413–34.

Mariani, John F. 1999. *The Encyclopedia of American Food and Drink*. New York: Lebhar-Friedman.

Marks and Spencer. 2014. "Forever Fish." September 12. Accessed May 15. http://corporate.marksandspencer.com/plan-a/find-out-more/about-our-initiatives/forever-fish.

McCright, Aaron M. 2010. "The Effects of Gender on Climate Change Knowledge and Concern in the American Public." *Population and Environment* 32 (1): 66–87.

McKinley, Andrew C., Laura Ryan, Melinda A. Coleman, Nathan A. Knott, Graeme Clark, Matthew D. Taylor, and Emma L. Johnston. 2011. "Putting Marine Sanctuaries into Context: A Comparison of Estuary Fish Assemblages over Multiple Levels of Protection and Modification." *Aquatic Conservation: Marine and Freshwater Ecosystems* 21 (7): 636–48.

McQueen, Craig. 2008. "The Unlikely Friends behind Scots Oyster Firm That Became International Hit." *Daily Record*, June 14.

McRobbie, Angela. 1982. "The Politics of Feminist Research: Between Talk, Text and Action." *Feminist Review* 12: 46–57.

Meinzen-Dick, Ruth, Chiara Kovarik, and Agnes R. Quisumbing. 2014. "Gender and Sustainability." *Annual Review of Environment and Resources* 39: 29–55.

Mentz, Steve. 2009a. *At the Bottom of Shakespeare's Ocean*. London: Continuum.

Mentz, Steve. 2009b. "Toward Blue Cultural Studies: The Sea, Maritime Culture, and Early Modern English Literature." *Literature Compass* 6 (5): 997–1013.

Mentz, Steve. 2012. "After Sustainability." *PMLA* 127 (3): 586–92.

Mezzadra, Sandro, and Brett Neilson. 2013. *Border as Method, or, the Multiplication of Labor*. Durham, NC: Duke University Press.

Miele, Mara. 2011. "The Taste of Happiness: Free Range Chicken." In "The New Borders of Consumption," special issue, *Environment and Planning A* 43: 2076–290.

Miele, Mara, and Adrian Evans. 2010. "When Foods Become Animals: Ruminations on Ethics and Responsibility in Care-Full Practices of Consumption." *Ethics, Place and Environment* 13 (2): 171–90.

Mighetto, Lisa. 2005. "Lisa Mighetto on Mermaids, the *Pacific Fisherman*, and the 'Romance of Salmon.'" *Environmental History* 10 (3): 532–37.

Mol, Annemarie. 2008. *The Logic of Care: Health and the Problem of Patient Choice*. London: Routledge.

Mol, Annemarie. 2013. "Mind Your Plate! The Ontonorms of Dutch Dieting." *Social Studies of Science* 43 (3): 397–416.

Morabia, Alfredo, and Anne Hardy. 2005. "The Pioneering Use of a Questionnaire to Investigate a Food Borne Disease Outbreak in Early 20th Century Britain." *Journal of Epidemiology and Community Health* 59 (2): 94–99.

"More Anchovies Please." 2011. *Green Solutions Magazine*. September 29. http://www.greensolutionsmag.com/?p=1933.

Morton, Timothy. 2010. "Here Comes Everything: The Promise of Object-Oriented Ontology." *Qui Parle: Critical Humanities and Social Sciences* 19 (2): 163–90.

Morton, Timothy. 2013. *Hyperobjects: Philosophy and Ecology after the End of the World*. Minneapolis: University of Minnesota Press.

Murray, Rupert. 2009. *The End of the Line: Imagine a World without Fish* [film]. London: Dogwoof Pictures.

Nadel-Klein, Jane. 2003. *Fishing for Heritage: Modernity and Loss along the Scottish Coast*. Oxford: Berg.

Nat Geo Wild. 2013. *Mission: Save the Ocean* [film].

Neis, Barbara. 2005. "Introduction." In *Changing Tides: Gender, Fisheries and Globalization*, edited by Barbara Neis, Marian Binkley, Siri Gerrard, and Maria C. Maneschy, 1–13. Newfoundland: ISER.

Newton, John. 2014. "Academic Pia Winberg, Chefs Mark Best and Victor Liong and Others Are Planting Seaweed on Menus." *The Australian*, August 16.

Nightingale, Andrea. 2006. "The Nature of Gender: Work, Gender, and Environment." *Environment and Planning D: Society and Space* 24 (2): 165–85.

Nightingale, Andrea. 2013. "Fishing for Nature: The Politics of Subjectivity and Emotion in Scottish In-Shore Fisheries Management." *Environment and Planning A* 45 (10): 2363–78.

Norton, Rictor. 1998. *My Dear Boy: Gay Love Letters through the Centuries*. San Francisco: Leyland.

NPR. 2003. "Ed Ricketts and the 'Dream' of Cannery Row." *Morning Edition*, May 7. http://www.npr.org/templates/story/story.php?storyId=1252560.

Ogden, Lesley Evans. 2013. "Aquaculture's Turquoise Revolution: Multitrophic Methods Bring Recycling to the Seas." *BioScience* 63 (9): 697–704.

Omega Protein. 2015. "Who We Are." Accessed March 2. http://omegaprotein.com/who-we-are/.

Páez, Ángel. 2011. "Malnutrition Has an Indigenous Face in Peru." Inter Press Service, January 18.

Pálsson, Gísli. 1994. "Enskilment at Sea." *Man (Journal of the Royal Anthropological Institute)* 29 (4): 901–27.

Pálsson, Gísli, and A. Helgason. 1995. "Figuring Fish and Measuring Men: The Individual Transferable Quota System in the Icelandic Cod Fishery." *Ocean and Coastal Management* 28 (1): 117–46.

Pálsson, Gísli, Bronislaw Szerszynski, Sverker Sörlin, John Marks, Bernard Avril, Carole Crumley, Heidi Hackmann, Poul Holm, John Ingram, Alan Kirman, Mercedes Pardo Buendía, and Rifka Weehuizen. 2013. "Reconceputalizing the 'Anthropos' in the Anthropocene: Integrating the Social Sciences and Humanities in Global Environment Change Research." *Environmental Science and Policy* 28: 3–13.

Pauly, Daniel. 1995. "Anecdotes and the Shifting Baseline Syndrome of Fisheries." *Trends in Ecology and Evolution* 10 (10): 430.

Pauly, Daniel. 2009. "Aquacalypse Now." *New Republic*, September 28.

Pauly, Daniel, and I. Tsukayama, eds. 1987. *The Peruvian Anchoveta and Its Upswelling Ecosystem: Three Decades of Change*. Callao, Peru: IMARPE.

Peters, Kimberley. 2010. "Future Promises for Contemporary Social and Cultural Geographies of the Sea." *Geography Compass* 4 (9): 1260–72.

Peters, Kimberley, and Philip Steinberg. 2015. "A Wet World: Rethinking Place, Territory and Time." *Society and Space*, April 28.

Pikitch, E., P. D. Boersma, I. L. Boyd, D. O. Conover, P. Cury, T. Essington, S. S. Heppell, E. D. Hourde, M. Mangel, D. Pauly, É. Plagányi, K. Sainsbury, and R. S. Steneck. 2012. *Little Fish, Big Impact: Managing a Crucial Link in Ocean Food Webs*. Washington, DC: Lenfest Ocean Program.

Plumwood, Val. 2007. "Human Exceptionalism and the Limitations of Animals: A Review of Raimond Gaita's *The Philosopher's Dog*." *Australian Humanities Review* 42.

Powell, Robin. 2009. "Natural Selection." *The Age*, October 20.

Power, Nicole G. 2000. "Women Processing Workers as Knowledgeable Resource Users." In *Linking Fishery People and Their Knowledge with Science and Management*, edited by Barbara Neis and Lawrence Felt, 183–203. St. Johns, Newfoundland: INSER.

Power, Nicole Gerarda. 2005. "The 'Modern Fisherman': Masculinity in Crisis or Resilient Masculinity?" *Canadian Women's Studies* 24 (4): 102–7.

Probyn, Elspeth. 1995. "Perverts by Choice: Towards an Ethics of Choosing." In *Feminism Beside Itself*, edited by Diane Elam and Robyn Wiegman, 261–82. New York: Routledge.

Probyn, Elspeth. 2000. *Carnal Appetites: FoodSexIdentities*. London: Routledge.

Probyn, Elspeth. 2004a. "Eating for a Living: A Rhizo-ethology of Bodies." In *Cultural Bodies: Ethnography and Theory*, edited by T. Thomas and J. Ahmed, 215–40. Oxford: Blackwell.

Probyn, Elspeth. 2004b. "Shame in the Habitus." In *Feminism after Bourdieu*, edited by Lisa Adkins and Beverley Skeggs, 224–48. Oxford: Blackwell.

Probyn, Elspeth. 2004c. "Thinking with Gut Feeling." *Public* 30: 101–12.

Probyn, Elspeth. 2005a. *Blush: Faces of Shame*. Minneapolis: University of Minnesota Press.

Probyn, Elspeth. 2005b. "Thinking Habits and the Ordering of Life." In *Ordinary Lifestyles*, edited by David Bell and Joanne Hallows, 243–54. Milton Keynes, U.K.: Open University Press.

Probyn, Elspeth. 2012. "Eating Roo: Of Things That Become Food." *New Formations* 74: 33–45.

Probyn, Elspeth. 2015. "Listening to Fish: More-Than-Human Politics of Food." In *Nonrepresentational Methodologies: Re-envisioning Research*, edited by Phillip Vannini, 72–88. New York: Routledge.

Puig de la Bellacasa, María. 2010. "Ethical Doing in Naturecultures." *Ethics, Place and Environment: A Journal of Philosophy and Geography* 13 (2): 151–69.

Pure Salmon Campaign. 2015. "Waste Contamination from Salmon Farms." Accessed March 5. http://www.puresalmon.org/waste_contamination.html.

Ransley, Jesse. 2005. "Boats Are for Boys: Queering Maritime Archaeology." *World Archaeology* 37 (4): 621–29.

Raschka, Christopher. 1998. *Arlene Sardine*. New York: Orchard.

Ray, Greg. 2013. "The Ocean Is Broken." *Sydney Morning Herald*, October 18.

Renton, Alex. 2006. "How Sushi Ate the World." *Guardian*, February 26.

Richards, Morgan. 2013. "Global Nature, Global Brand: BBC Earth and David Attenborough's Landmark Wildlife Series." *Media International Australia* 146: 143–54.

Roe, Emma. 2006. "Things Becoming Food and the Embodied, Material Practices of an Organic Food Consumer." *Sociologica Ruralis* 46 (2): 104–21.

Rojas-Ruiz, Jorge. 2014. "Aquaculture: Enabling Food Security, Oceanic Sustainability and Economic Growth in the Future." *Hunger and Undernutrition Blog*, World Bank, April 24. http://www.hunger-undernutrition.org/blog/world-bank/.

Romano, Stefania, Vincenzo Esposito, Claudio Fonda, Anna Russo, and Roberto Grassi. 2006. "Beyond the Myth: The Mermaid Syndrome from Homerus to Andersen: A Tribute to Hans Christian Andersen's Bicentennial of Birth." *European Journal of Radiology* 58 (2): 252–59.

Rowse, Tim. 2002. *White Flour, White Power: From Rations to Citizenship in Central Australia*. Cambridge: Cambridge University Press.

Sanderson, Rosemary. 2008. *The Herring Lassies: Following the Herring*. Banffshire, Scotland: Bard.

SBS. 2014. "What's the Catch? Prawns" [online document]. November 3. http://www.sbs.com.au/programs/whats-the-catch/prawns.

Scottish Tartans Authority. 2014. "Highland Clearances." Accessed October 27. http://www.tartansauthority.com/resources/the-highland-clearances/.

Sedgwick, Eve Kosofsky. 1993a. "Queer and Now." In *Tendencies*, 1–22. Durham, NC: Duke University Press.

Sedgwick, Eve Kosofsky. 1993b. *Tendencies*. Durham, NC: Duke University Press.

Shepherd, C. J., and A. J. Jackson. 2012. "Global Fishmeal and Fish Oil Supply—Inputs, Outputs, and Markets." Paper presented at the Sixth World Fisheries Congress, Edinburgh, May 9.

Simmel, Georg. 1994. "The Sociology and the Meal." *Food and Foodwayus* 5 (4): 345–50.

Singer, Peter. 1990. *Animal Liberation*. New York: HarperCollins.

S J Noble Trust. 2014. "About Us." Accessed October 27. http://www.sjnobletrust.com/about.php.

Skeggs, Beverley. 2004. "Exchange, Value and Affect: Bourdieu and 'the Self.'" In *Feminism after Bourdieu*, edited by Lisa Adkins and Beverley Skeggs, 75–96. Oxford: Blackwell.

Slocum, Rachel. 2008. "Thinking Race through Corporeal Feminist Theory: Divisions and Intimacies at the Minneapolis Farmers' Market." *Social and Cultural Geography* 9 (8): 849–69.

Smith, Lewis. 2011. "Sustainable Fish Customers 'Duped' by Marine Stewardship Council." *Guardian*, January 6.

Soeters, Karen, and Gertjan Zwanikken. 2008. *Meat the Truth* [film]. Netherlands: Nicolaas G. Pierson Foundation.

Steele, E. N. 1964. *The Immigrant Oyster (Ostrea Gigas): Now Known as the Pacific Oyster*. Olympia: Warren's Quick Print.

Steinbeck, John. 1945. *Cannery Row*. New York: Viking.

Steinberg, Philip E. 1999a. "The Maritime Mystique: Sustainable Development, Capital Mobility, and Nostalgia in the World Ocean." *Environment and Planning D: Society and Space* 17 (4): 403–26.

Steinberg, Philip E. 1999b. "Navigating to Multiple Horizons: Toward a Geography of Ocean-Space." *Professional Geographer* 51 (3): 366–75.

Stott, Rebecca. 2000. "Through a Glass Darkly: Aquarium Colonies and Nineteenth-Century Narratives of Marine Monstrosity." *Gothic Studies* 2 (3): 305–27.

Stott, Rebecca. 2004. *Oyster*. London: Reaktion.

Strathern, Marilyn. 2012. "Eating (and Feeding)." *Cambridge Journal of Anthropology* 30 (2): 1–14.

Sydney Fish Market. 2015. "Our Company." Accessed June 1. http://www.sydneyfishmarket.com.au/our-company/our-company.

Sydney Living Museums. 2012. "Of Oyster Shells and Shelly Mortar." The Cook and the Curator, November 22. http://blogs.hht.net.au/cook/oyster-shells/.

Tamburri, Mario N., Richard K. Zimmer, and Cheryl Ann Zimmer. 2007. "Mechanisms Reconciling Gregarious Larval Settlement with Adult Cannibalism." *Ecological Monographs* 77 (2): 255–68.

Tavee, Tom, and H. Bruce Franklin. 2001. "The Most Important Fish in the Sea." *Discover*, September 1.

Thompson, Paul. 1985. "Women in the Fishing: The Roots of Power between the Sexes." *Comparative Studies in Society and History* 27 (1): 3–32.

Thompson, Paul, Tony Wailey, and Trevor Lummis. 1983. *Living the Fishing*. London: Routledge and Kegan Paul.

Thrift, Nigel. 2008. *Non-representational Theory: Space, Politics, Affect*. London: Routledge.

Throsby, Karen. 2013. "'If I Go In Like a Cranky Sea Lion, I Come Out Like a Smiling Dolphin': Marathon Swimming and the Unexpected Pleasures of Being a Body in Water." *Feminist Review* 103: 5–22.

"Time to Shell Out! Oysters Make a Return to Oystermouth a Century after Being Wiped Out by Pollution." 2014. *Daily Mail*, October 24.

Tinellis, Claudine. 2014. *Coastal Chef: The Culinary Art of Seaweed and Algae in the 21st Century*. Ulladulla, Australia: Harbour.

Transparency Market Research. 2015. "Booming Aquaculture Industry to Boost Demand from Global Fish Oil Market." March 25. http://www.transparencymarketresearch.com/article/fish-oil.htm.

Tsing, Anna Lowenhaupt. 2014. "Strathern beyond the Human: Testimony of a Spore." *Theory, Culture and Society* 31 (2–3): 221–41.

Wacquant, Loïc. 2004. "Habitus." In *International Encyclopedia of Economic Sociology*, edited by Jens Beckert and Milan Zafirovski, 315–19. London: Routledge.

Wacquant, Loïc. 2014. "Homines in Extremis: What Fighting Scholars Teach Us about Habitus." In *Fighting Scholars: Habitus and Ethnographies of Martial Arts and Combat Sports*, edited by Raúl Sánchez García and Dale C. Spencer, 193–200. London: Anthem.

Waters, Sarah. 1998. *Tipping the Velvet*. London: Penguin.

Watson, Reg. 2015. "Close Two-Thirds of the Ocean to Make Fishing Better and Fairer." *The Conversation*, February 18.

Watson, Reg, and Daniel Pauly. 2001. "Systematic Distortions in World Fisheries Catch Trends." *Nature* 414: 534–36.

Weheliye, Alexander G. 2014. *Habeas Viscus: Racializing Assemblages, Biopolitics, and Black Feminist Theories of the Human*. Durham, NC: Duke University Press.

Whatmore, Sarah. 2004. "Humanism's Excess: Some Thoughts on the 'Post-human/ist' Agenda." *Environment and Planning A* 36 (8): 1360–63.

Whatmore, Sarah. 2006. "Materialist Returns: Practising Cultural Geography in and for a More-Than-Human World." *Cultural Geographies* 13 (4): 600–609.

Williams, Meryl. 2010. *Addressing Human Capital Development and Gender Issues in the Aquaculture Sector*. Rome: FAO.

Williams, Raymond. 1989. "Culture Is Ordinary." In *Resources of Hope: Culture, Democracy, Socialism*, 3–14. London: Verso.

Wills, Wendy, Kathryn Backett-Milburn, Mei-Li Roberts, and Julia Lawton. 2011. "The Framing of Social Class Distinctions through Family Food and Eating Practices." *Sociological Review* 59 (4): 725–40.

Wintersteen, Kristin. 2012. "Sustainable Gastronomy: A Market-Based Approach to Improving Environmental Sustainability in the Peruvian Anchoveta Fishery." In *Environmental Leadership: A Reference Handbook*, vol. 2, edited by D. Gallagher, 626–634. Thousand Oaks, CA: Sage.

Wishart, Ruth. 2008. "For One Firm's Workers, the Oyster's Their World." *Herald Scotland*, June 11.

WorldFish. 2014. "Giving Women a Voice in Ghana's Coastal Resource Management." November 6. http://www.worldfishcenter.org/content/giving-women-voice-ghana%E2%80%99s-coastal-resource-management.

World Ocean Review. 2015. "The Global Hunt for Fish." Accessed April 9. http://worldoceanreview.com/en/wor-2/fisheries/state-of-fisheries-worldwide/.

WWF. 2006. "Australians Encouraged to Become Futuremakers." April 3. http://www.wwf.org.au/?2247/Australians-encouraged-to-become-Futuremakers.

WWF. 2015a. "Blackmores." Accessed February 16. http://www.wwf.org.au/about_us/working_with_business/strategic_partnerships/blackmores/.

WWF. 2015b. "Fishmeal and Fish Oil." Accessed February 16. http://www.worldwildlife.org/industries/fishmeal-and-fish-oil.

Yates-Doerr, Emily. 2012. "The Weight of the Self: Care and Compassion in Guatemalan Dietary Choices." *Medical Anthropology Quarterly* 26 (1): 136–58.

Yeoman, Rose. 2014. "Dished Up: Rediscovering Seaweed Links in Food Chain." Fisheries Research and Development Corporation, March. http://frdc.com.au/knowledge/publications/fish/Pages/22-1_articles/20-dished-up-rediscovering-seaweed.aspx.

Yonge, C. M. 1960. *Oysters*. London: Collins.

Yusoff, Kathryn. 2013. "Insensible Worlds: Postrelational Ethics, Indeterminacy and the (K)nots of Relating." *Environment and Planning D: Society and Space* 31 (2): 208–26.

INDEX

Abaka Edu, Emelia, 125
abalone, 30, 159
Abbott, Tony, 30
Aberdeen, 34, 111
Aboriginal people, 1, 61, 63, 79; under colonial occupation, 63; and fish traps, 84; and middens, 55, 167n4; and nutrition, 3; in trade, 124
Abrahamsson, Sebastian, et al., 11, 52, 136–38, 157
activism, 27–28, 33, 51, 124
Acurio, Gaston, 144–46
affect (in Deleuzian sense), 11, 15–16, 27, 31, 34, 42–43, 46, 64, 110, 168nn2–3; habitus, 36–38, 44; in Latour, 59
Africa: fishing in, 5, 124–25, 157; migratory fish and, 83; NGOs in, 143; West Africa, 96, 138, 157
African Americans, 109
Agarawal, Bina, 107
agriculture, 4, 15, 25, 47, 151, 154, 163; Food and Agriculture Organization (FAO), 8–9, 84. *See also* farming
air pollution. *See* pollution
Alaimo, Stacy, 32–33, 46
alcohol, 58, 120
algae, 13, 69; aquaculture and, 73, 156–57; blooms of, 140; eating of, 130, 155; within ecosystems, 24, 53; mercury in, 32

Alice in Wonderland, 55
"alien" water environments, 26, 32, 79
Allewaert, M., and Michael Ziser, 33–34
Alzheimer's prevention, 135
amino acid, 53
anchovies, 19, 140, 143, 149; eating, 162–63; ecosystems and, 130; as forage fish, 139; paste, 147; relations of, 148–49. *See also* Peruvian anchoveta
ancient Romans, 52–53, 147, 151
Andersen, Hans Christian, 103, 112–13
angasi oysters (*Ostrea angasi*), 62, 71
Animal Liberation, 25
Antarctic, 30, 61, 162
Anthropocene, 12, 51, 104
antibiotics, 81
aphrodisiacs, 53
"Aquacalypse," 13, 125
aquaculture, 7, 95, 139; education programs in, 69–70, 75; models of, 25, 151–52, 154, 157, 163; multitrophic, 73, 130, 153; soybean meal in, 156; sustainability of, 13, 150, 155; technology in, 74. *See also* integrated marine trophic aquaculture (IMTA)
aquarium, 97, 102
Arctic, 32, 104
Arlene Sardine, 133
Arnhem Land, 84
"athwart," 18–19, 21, 37, 55, 83, 126

"attunement," 67–68, 70, 131
Australia: exports, 96; fish, 20, 83–85, 133; fishing industry, 29–30, 96, 162; food culture, 3, 97, 134–36, 161; horizon, 40; natural environment, 4, 150; ocean currents, 1; oysters and, 55, 61–63, 68, 74–75; pearling, 167n1; people, 27, 39, 86, 124; research, 6, 17; tuna industry, 87–95, 99. *See also* Aboriginal people; South Australia

Bachelard, Gaston, 41
bacteria, 24, 74, 139, 149
bait, 45, 91, 114–15, 141; live bait fishing, 90
Banffshire Maritime Heritage Association, 115
Barad, Karen, 109
Barclay, Kate, 168n3; and Charlotte Epstein, 86, 167n1
barramundi, 70, 84
Barthes, Roland, 41
Bear, Christopher, and Sally Eden, 26, 111
beef, 7, 86, 156
Bennett, Jane, 109, 136, 138, 149
BHP, 68
big fish, 24, 32, 99, 137, 157, 162
Big Pharma, 138
big predators. *See* predators
Billingsgate Fish Market, 19–20
bioluminescence, 79
Birke, Lynda, 109
bivalves, 13, 53–54, 60, 66, 71, 130, 162
Blackmores, 134–35
bluebottle jellyfish, 79
bluefin tuna, 12, 14, 75, 81; as endangered species, 77–78, 95, 168n4; farming, 80, 92; fishing, 88–90; and globalization, 15, 87; ITQs of, 91–92, 166n3; market, 19, 85–87, 95–98; migration of, 83; South Australian, 4; technology and, 84; tourism and, 81–83. *See also* tuna
Blue Rage, 78
Blumenthal, Heston, 104
borders, 129
Boston Bay, 80–81, 92, 97
Bourdieu, Pierre, 35–36, 58–59
BP, 110
Branson, Richard, 28, 44

bread, 6, 59, 135–36
brisling, 132–33
British Commonwealth, 63, 167n6
Bruce, Stanley, 115
Bryld, Mette, and Nina Lykke, 109
BST, 70
Buckie Heritage Centre, 120–21
Burton, Valerie, 117
Butler, Judith, 108, 109, 113

Cairndow, 62, 64, 67
calcium, 53, 59
calenture, 27
California, 19, 27, 60, 142, 149, 156; canneries, 131, 133
Callon, Michel, 67
Canada, 13, 107; coastline, 85; fisheries, 13, 122–23; ITQs, 4; politics, 124
Canadian Broadcasting Company (CBC), 123
cancer, 32
canneries, 90, 131–33
Cannery Row, 130–31, 133
carnivorous fish, 24, 120, 156
Carroll, Lewis, 55–58
Carson, Rachel, 47
Casanova, 53
caviar, 97
celebrities, 27–28, 165n1
CGIAR Consortium, 106
chef, 28, 71, 144–46, 155
Chesapeake Bay, 53, 139
Chile, 139, 154
China, 9, 13, 62; aquaculture in, 153–54; fish farming in, 143–44; manufacturing in, 1; tourists from, 159
Chopin, Thierry, 152–54, 168n3
Church, 115
clapotis, 23, 46
class (social), 12, 105, 108, 166n2; and discrimination, 3; and education, 59; and identity, 96; middle, 44, 107, 134, 147, 152; and taste, 58–61, 75, 133; upper, 115, 131, 151; working, 36, 120, 135
climate change, 11–12, 33, 105, 119. *See also* ecofeminism
Club des Millionaires, 131–32
Clupeidae family, 115, 139

cobles (boats), 114
cod, 12–13, 104, 114, 122–23
Coffin Bay, 62, 65, 161
Coleridge, Samuel Taylor, 23, 38, 163
colonization, 20, 96, 104; colonizers, 3, 139; trade routes under, 5; white death, 3
commercial fishing, 2, 29–30, 91, 120. *See also* fishing industry
conservation: advocates of, 8, 32, 73, 78, 151; legislation and policy for, 92, 94, 124, 168n4; practices of, 20, 84, 106
contamination, 81, 137
Convention on International Trade in Endangered Species (CITES), 96
Cook, Captain James, 102
copper, 59, 68
cow, 25; beef, 7, 86, 156
Cowell, 68–69, 74, 80
crab, 131; meat, 87; mud crab, 160
Crassostrea gigas. *See* Pacific oysters
Croatia, 15, 97; migrants from, 88–89, 95; tuna ranching in, 92
Cromwell, Oliver, 114
cultural politics, 23, 27, 30, 48
culture shock, 26
curing, 114–15, 117, 119, 145
cutlery. *See* fork

Danson, Ted, 27
Darwinian evolution, 113
Davis, Dona Lee, 13, 123, 127
"dead zones," 24, 140
death, 12, 36, 64, 102, 112, 119–20, 122, 156
deep-sea mining, 31
de Lauretis, Teresa, 108, 109, 125
Deleuze, Gilles, 25; ethology, 14–16, 46–47; and Félix Guattari, 14
demersal fish, 119, 137
Denmark, 139
depression, 6, 136
Depression, the, 133
didgeridoo, 1
disease, 57–58, 81, 139, 149
Disney, 103, 113
documentary, 27–30, 43
Drawing the Line (documentary), 29–30
dredging, 53, 62, 71, 75

drought, 6–7
drugs, 129
Duarte, Carlos, 7, 27, 150

earthquake, 27
eating: cultural identity and, 6–7, 40, 58, 134; and environmental impact, 2, 10, 97, 130, 151, 165n1; and health, 138; oysters, 11, 50–51, 55, 59; pescetarianism, 25; practices of, 75, 147; relations and, 11, 51–52, 152. *See also* class (social); food
Eating Body, 11, 166n5
eco-consumer, 105
ecofeminism, 106–7
ecology: of Franklin Harbour, 70; marine, 32, 66, 155; "spinster," 47; theory, 55
Economist, Big Mac index of, 97
ecosystems, 11, 25, 142, 147, 152–53, 155–57; changing, 2, 24, 51, 144; damage to, 7, 29, 43, 142; menhaden and, 139–40; oysters and, 53, 68; social, 34
eggs, 24; chicken, 59, 136–37; oyster, 53
El Niño, 142–43
employee ownership, 64, 68, 72–74
endangered species, 20, 77–78, 83, 93, 96
End of the Line, The (documentary), 27–28, 31, 96,
England, 62, 71, 118, 149; people of, 39, 59; Victorian, 54, 57–58, 102–3
enskilment, 42–43, 83, 94
Ensor, Sarah, 47
entrepreneur, 28, 124, 132, 142
environment: changes in, 52, 105–7; contamination of, 137, 151; issues of, 33, 110, 144, 152, 157; relations of, 40; sustainability of, 143, 145, 147
epidemiology, 57, 137
Erdrich, Louise, 129, 158
ethics: in farming, 7; of food, 4, 6, 12, 19–20, 77; and relations, 14, 46, 163
ethology, 10, 14–16, 19, 46; rhizo-ethology, 14, 50
EU, 28, 138, 157
Evans, Adrian, and Mara Miele, 50, 77, 166n5
Exclusive Economic Zones, 84, 138
extinction, 53, 71, 75, 78, 95
Eyre Peninsula, 6, 61, 68–70, 81, 84

Faier, Lieba, and Lisa Rofel, 126
farmers, 4, 62–64, 81, 87, 99, 105, 123, 149; communities, 6, 69, 92, 97; markets, 107
farming, 62–65, 80, 139, 155, 166n5; anti-fish, 152; Chinese, 143–44; fish, 7, 19–20, 73–74, 80–81, 87, 89, 93, 98, 100, 130, 140, 152, 157; oyster, 12, 53, 69–70, 162, 167n1; pig, 149, 156; poultry, 156; salmon, 150–51; shrimp, 5; tuna, 95, 100. *See also* agriculture
Fearnley-Whittingstall, Hugh, 28
feminism: and cultural studies, 37; ecofeminism, 106–7; and ethics, 12, 16, 46, 106, 131; and fisheries, 123; and gender, 104, 109; and labor, 46; and the posthuman, 32; and queer theory, 17–18, 126; relations and, 18, 127, 168n3; and sociologists, 13
feminists, 104, 107–9, 110
feminization, 123
fertilizer, 5, 142, 146, 152–54
Festival de Films Pêcheurs du Monde, 143
fever. *See* calenture
Filipinos, 120
First Nations peoples, 139
fish: eggs, 24, 97; farming, 7, 19, 63, 74, 81, 144, 150–52, 157; fish-as-food, 4–5; global scale of, 52; guts, 113, 117–18; industrial processing of, 134, 139, 142; traps, 84. *See also* fisheries; fish meal; fish oil; little fish
Fisher, M. F. K., 12, 54–55, 60
fisheries, 27, 134, 143, 145, 146, 150, 162; Atlantic, 122; management of, 94, 122, 124; scientists at, 30, 34
fishermen, 5, 8, 18, 44, 62, 78, 93, 111, 118–20, 123, 133, 163
fishers, 143, 24, 29–31, 34, 42–44, 80, 85, 91–92; recreational, 2, 85
fishertouns, 114. *See also* Scotland
Fish Fight, 28
Fishing for Heritage, 113
fishing industry, 29; licenses, 146; technology, 92
fishing village, 34, 114, 160
FishLove, 78, 165n1
fish markets: Billingsgate, 19–20; Sydney, 159–62; Tsukiji, 19, 78, 83, 85–86, 96
fish meal, 29, 134, 129, 143–44

fish oil, 5, 29, 133–40, 145, 154, 157
fishwives, 18, 100, 113, 117. *See also* fish-women
fish-women, 102, 111–12, 126; of Scotland, 114–15, 117
Flower of Scotland, 62
food: animal protein in, 7; animals and, 5, 55, 85–86, 130–34; economy and, 43, 144, 146; eroticism, 53; ethics of, 4, 6, 12, 19–20, 77; global population and, 24; industry, 27, 50, 70, 80, 96, 139, 150, 153–55; in literature, 60; and the ocean, 62, 65; politics of, 2, 6, 11, 19, 23, 25, 27, 86, 106; security, 86, 147; slow food, 143; socio-economic status and, 115, 133; super foods, 133; waste, 149, 151; women and, 104, 125. *See also* oysters; protein; taste
Food and Agriculture Organization (FAO), 8–9, 84
food chain, 5, 24, 32, 96, 130, 136, 139–40, 144, 156
food web, 10, 24, 140
Footdee, 34
fork, 10, 35, 47, 165n2; as weapon, 29, 31, 44, 51
Foucault, Michel, 40, 107, 109
France, 58, 143
Franklin, H. Bruce, 139
Franklin Harbour, 69–71, 139
Fukushima, 27

Gallic saying, 71
Gallup data, 105
Gatens, Moira, 16, 46–47
gender: discourse of, 37, 101–2; and environment, 104–7, 110–11; equality of, 104, 108; in fishing, 13, 18, 100, 117–18, 123, 125, 150; fluidity of in oysters, 54, 75; identity and, 9, 113; relations of, 10, 12, 16, 60, 94, 109–10, 126. *See also* feminism; mermaids; sexuality
GenderAquaFish, 150
Genderfish, 18
Gender Trouble, 108
geography, 14, 19, 44, 122, 137, 151; of animals, 26; cultural, 50; and feminism, 110; of the ocean, 39
geopolitics, 4, 19, 105, 157

gibbing, 117
Gibson-Graham, J. K., 110, 131
Glasgow, 49, 62, 152
globalization, 15, 37, 83, 86–87, 124
Global North, 4–5, 105
Global South, 18, 92, 105, 124–25, 143–44, 157
GLOBEC, 150
gold, 62
Goodman, David, 83
Goodman, Michael K., 43–44
GPS, 146
Grand Banks of Newfoundland, 122
Great Australian Bight, 83, 90
greed, 15, 58, 63, 93, 112, 145
Greek mythology. *See* mermaids; sirens
Greenberg, Paul, 53, 81, 87, 140
Greene King, 71, 167n8
Greenpeace, 27–28
guano, 142
Guattari, Félix, and Gilles Deleuze, 14

habitat, 6, 52, 61, 67, 70, 137, 167n7
habitus, 11, 35–38, 43–45, 47, 59, 63
haddock, 114
hake, 138
Haraway, Donna, 110–11, 127, 168n3
Hardin, Garrett, 44, 93–94
Hawai'i, 69
Haynes-Conroy, Alison, and Jessica Haynes-Conroy, 60
Hayward, Eva, 109, 130
health supplements. *See* supplements
heavy metals, 2, 15
Heise, Ursula, 105
Helmreich, Stefan, 15, 18, 37, 82, 126, 129
Hennion, Antoine, 51, 59
hermeneutics, 40
herring, 12–14, 139; fishing of, 114; industry, 115, 119–20, 125; lassies, 19, 113, 117–18; quines, 100
hierarchical constructs: in colonialism, 63; in food, 27, 58; in species, 10
Highlands, 62–63, 72, 113, 147
Homer, 99, 91, 102
homing, 99, 111, 113, 152
homosexuality, 113

Hong Kong, 64, 67, 74
hooks, 114
horizon, 40, 42, 46, 80
hormones, 53
horse mackerel, 85
Howard, Penny, 122
human-fish entanglements, 5, 12, 42, 112, 156
human-fish relations, 8, 10, 15, 19, 26, 47, 158, 163
Humbolt Current, 141
Hurricane Sandy, 53
Huxley, Thomas H., 27, 29, 51, 122
"hyperobjects." *See* Morton, Timothy
hypoxia, 24, 34, 53, 140. *See also* oxygen

Iceland, 4, 26, 42
Idaho, 69
Indian Ocean, 83
individual transferrable quotas (ITQs), 4, 91–92, 94, 124
Indonesia, 92, 146, 156; Java, 83, 90
Ingold, Tim, 66
integrated marine trophic aquaculture (IMTA), 13, 153–55, 163
International Fishmeal and Fish Oil Association (IFFO), 139, 141, 145–46
International Food Policy Research Institute, 106
Inuit, 32
Inveraray, 72–73
Inverness, 114. *See also* Scotland
invertebrae, 24
iodine, 59
Ireland, 115; fishers of, 62, 122
iron, 53; ore, 68
Issenberg, Sasha, 86, 96

Jacobsen, Rowan, 52
Japan: earthquake, 27; fishing, 90, 92; food habits, 70; tuna consumption, 85–87, 92, 95–97
Java, 83, 90
jellyfish, 2, 24, 26, 79, 109, 140
John West, 9, 90, 93, 104

Kawafuku Restaurant, 86
Kincaid, Trevor, 54

Kindai University, 95
Kinsey, Alfred, 112–13
kippers, 63, 72
klondykers, 119
Kolega, Rick, 94
Krummer, Corby, 133

LA Times, 145
Latour, Bruno, 59–60, 67, 70
Lee, Michael Parrish, 55
Le Sann, Alain, 124, 143–44
Libya, 96
Lima, 144
limpets, 114
little fish, 13, 129–30, 133, 137, 140–43, 146, 149–50, 157, 162
Little Fish, Big Impact, 142
Little Mermaid, The, 103, 113
lobster, 24–25, 30, 62, 122, 154, 160, 162, 166n3
Loch Fyne Oysters, 14, 49, 62–65, 67–68, 70–71, 167n7; takeover, 72–75
London, 57, 62, 64, 78; Billingsgate Fish Market, 19–20
Lorimer, Hayden, 16
Lossiemouth, 111–12, 114, 119. *See also* Scotland
Lucilius, Gaius, 52
Lucius Junius Moderatus Columella, 151
Lukin, Dinko, 88–90, 92–93, 95, 97–98

MacArthur, General Douglas, 86
MacFadyen, Ivan, 27
mackerel, 20, 85
magnesium, 59
Maine, 24
Majluf, Patricia, 143–48, 151
Makassar, 124
malnutrition, 34, 144
mammals, 25
Mansfield, Becky, 137–39, 150–51, 157
Marine and Maritime Research Festival, 145
marine biologists, 27, 63, 103, 131, 143, 145
marine ecologists, 124
marine protected area (MPA), 29, 30,
marine science, 10, 24, 51, 70, 101, 125, 131
marine scientists, 49, 101, 131

Marine Stewardship Council (MSC), 28, 162
market economies, 11–12, 67, 119; American, 85, 87; export, 112, 144; Japanese, 96
marriage, 89–100, 103, 117, 119–20, 123, 160, 167n5
Mauritania, 143–44
McCright, Aaron, 105, 107
Meat the Truth, 28
menhaden, 13, 130, 139–41, 156
Mentz, Steve, 31–32, 46, 79
mercury poisoning, 32. *See also* methylmercury
mermaids, 12, 101, 113, 125–26, 155; in mythology, 102; song of, 103; symbolism of, 103
"metabolic intimacy," 11, 130, 148–49, 157–58
metabolization, 137, 149
methylmercury, 137
microplastics, 16
middle class. *See* class (social)
Miele, Mara, 136; and Adrian Evans, 50, 77, 166n5
Mighetto, Lisa, 103
Milton, Kay, 43
mining, 5–6, 32, 34, 62, 68; deep-sea, 31
misogyny, 47, 117. *See also* gender; sexism
Mission: Save the Ocean, 28
Mitsubishi, 95
Mol, Annemarie, 11, 52, 61, 75, 77, 104, 136, 148–49, 165n6, 166n5. *See also* Abrahamsson, Sebastian, et al.
mollusks, 12, 51, 53–54, 61
Montreal, 123, 131–32
"more-than-human," 10, 12, 15–16, 17, 19, 31, 33, 37–38, 43–44, 51, 127, 131, 142, 163, 166n5; in ecology, 75, 105; fish-women, 102, 113; and gender, 110, 114, 126; and metabolic intimacy, 148–49
Morton, Timothy, 104–5, 110
mud crab, 160
Mulroney, Brian, 124
multinationals: mining, 68; oil companies, 34
Murray River, 6
mussels, 24, 73, 114, 154, 157. *See also* oysters

Nadel-Klein, Jane, 19, 113, 117–18
Nanuo Aboriginal mob. *See* Aboriginal people

Neis, Barbara, 13, 18, 123–24
Netherlands, 28
nets, 85, 92, 94, 113–15, 132, 163; purse seine, 84, 92, 119, 133, 142
Newfoundland, 13, 122–24
New Yorkers, 53
New Zealand, 83, 92, 161
NGOs, 11, 18, 104–5, 107, 143, 150
Nicolaas G. Pierson Foundation, 28
Nightingale, Andrea, 43–44, 106–7, 110, 122
Nile, 20
nitrogen, 15, 53, 142, 151, 155
Nobu, 96–97
nonhuman, the, 10, 109–10
nonhuman bodies, 36, 60
nonhuman ecology, 25, 32, 68
nonhuman relations, 25, 42, 54–55, 110, 130, 163
nonrepresentational methodologies, 16
North America, 118; east coast of, 19, 85, 139; food consumption in, 96; oysters in, 53; Pacific Northwest, 152; tuna in, 96
Northern Territory (Australia), 3
Norton, Rictor, 112–13
Nova Scotia, 24
nudibranches, 130
nutrition, 3, 115, 133, 136, 147, 153, 156–57; malnutrition, 34, 144

ocean, 7, 9, 15; conceptualizing, 16, 18, 32, 39, 41; "dead zones," 24, 140; destruction of, 29, 31, 81, 162; "eating the ocean," 7, 77, 130, 150, 158, 163; as habitat, 137; Indian, 83; materiality of, 37; oceanic relations, 10, 27, 43, 47–48, 51, 77–78; Pacific, 1, 86, 125, 147, 151; power of, 11, 20; sustainability of, 28, 39; trophic cascading, 24. *See also* horizon
ocean pens, 95
Odyssey, The, 102
oil industry, 83
oil rig, 34
oil spill, 16, 110
Olympics, 90
Omega Protein, 139–40
omega-3, 135–40, 147, 156–57
One Steel, 68

Organization for the Promotion of Responsible Tuna Fisheries (OPRT), 91
Ostrea angasi, 62, 71
Ostrea edulis, 63–64, 71, 74. *See also* Loch Fyne Oysters
overfishing, 4, 29, 58, 78, 119, 144–46. *See also* sustainability
overpopulation, 7, 25, 93, 163
oxygen, 2, 24, 53, 83, 151, 155
Oyster. *See* Stott, Rebecca
oysters: in Aboriginal diet, 55; as bodies, 15, 25; and disease, 57; eating, 52, 59, 74, 77, 160, 163; farming industry of, 62–65, 68–71; as filters, 53, 67, 140; gender and, 61; overconsumption of, 58; predators of, 54; as queer, 53–55; relations with humans, 49–50, 75; shells of, 52; social class and, 59–613. *See also* Loch Fyne Oysters; Pacific oysters

Pacific Ocean, 1, 86, 125, 147, 151
Pacific oysters, 12, 62–63, 65, 71, 74–75, 161, 167n1
Pálsson, Gisli, 26, 40, 42–43, 83, 94
Party for Animals, 28
"pastoral nostalgia," 31
Pauly, Daniel, 9, 13, 27, 124–25, 141
pearl shell, 124
Peru, 19, 134, 139, 141–42, 145–47
Peruvian anchoveta, 140; and class, 147; industry of, 141–43, 145–46
pescetarianism, 25
pesticide, 2, 7, 32, 65, 154
Peterhead, 114, 120
phytoplankton, 13, 24, 53, 130, 137, 139–40, 154
pigs, 5, 140, 142, 149, 156
pilchard, 81–82, 90, 115
pirates, 31
plankton, 115, 139. *See also* phytoplankton; zooplankton
Plato, 31
politics of food, 2, 6, 11, 19, 23, 25, 27, 86, 106
pollution, 53, 57, 75, 81, 144
polyculture, 13, 151–52
poor, the, 9, 144; diets of, 2, 58–59, 115, 133, 147, 157

population: fish, 8–9, 78, 122; human, 3, 7, 25, 32, 62, 86, 93, 138, 143, 163; oyster, 53; rural, 62, 149; seabird, 142; shark, 80; shrinking, 69, 123. *See also* overpopulation
Port Lincoln, 79–82, 86–90, 92, 94–95, 97, 167n2
posthuman, the, 32, 109–10
postmodernism, 109
post-structuralism, 109
"post sustainable," the, 31–32, 46
poverty, 64, 105
Power, Nicole Gerada, 123–24
prawns, 24, 81, 138, 147, 161–62
predators, 2, 24, 32, 52, 66, 122, 130, 137–38, 140, 157
pregnancy, 117, 137–38
Pret a Manger, 97
Pristine Oysters, 71
privatization, 94, 124, 142, 166n3
progesterone, 53
protein, 7, 13, 25, 73, 124, 138–40, 142, 156, 162; affordable, 59, 144
Puglisi, Joe, 90, 94; family, 88, 99
purse seine, 84, 92, 119, 133, 142

queer ontology, 37, 126
queer relations, 54, 113; fish, 10, 12, 100
queer theory, 109
quines, 13–14, 100, 113, 115, 117, 119, 126
quota. *See* individual transferrable quotas (ITQs)

racism, 99, 150
Ransley, Jesse, 104
Raschka, Christopher, 133
raw fish, 86–87, 91, 97. *See also* sushi
raw oysters, 59
redundancy, 13
refugees, 129
relations, 42–46, 60, 67, 79, 109, 138. *See also* subjectivity
restaurants, 49, 71, 89; Peruvian, 144, 147; sushi, 78, 85–87; top, 64, 97, 159
"rhizo-ethology," 14, 50
Ricketts, Ed, 131
rip current, 20
Rofel, Lisa, and Lieba Faier, 126

Roman Empire. *See* ancient Romans
Round House, The, 129
runoff, 32, 154
rural depopulation, 69, 123

sailors, 26, 102–3, 163
salmon, 71; farming of, 7, 81, 104, 150–51, 156; fishing of, 114; homing, 111; wild, 4, 152. *See also* aquaculture; contamination
salting, 115, 117, 145
salt water, 32
Sanderson, Rosemary, 115, 117
Šantić, Tony, 88, 94
sardines, 13, 130, 139, 143, 145; affordability of, 20; Californian, 19, 149, 156; canning industry, 131–34; eating, 162; in food web, 81, 115, 139, 162; nutritional value of, 137. *See also* little fish
scallops, 24, 157
Scotland, 12, 19, 49, 62–63, 68; fishers in, 24, 43–44, 111, 120, 122; fisher-women in, 114–18; salmon farms in, 151–52
Scottish Salmon Company, 72–73
Scottish Seafood Investments, 72
seabirds, 38, 89, 142
sea cucumber, 13, 73, 130, 154, 157
sea grass, 69
seagull, 45, 159
sea lions, 80, 141, 143
seals, 80, 143, 166n6
Sea Shepherd Conservation Society, 78
seasickness, 26, 42, 79, 83
Sea the Truth (documentary), 28–29, 31, 43–44
seaweed, 151, 153–56
sea worms, 114
Sedgwick, Eve K., 18, 37, 83, 109, 126
SEKOL, 94
selenium, 53
sewage, 57–58
sex, 53–54, 60, 108, 113. *See also* gender; Kinsey, Alfred
sex change. *See* oysters
sexism, 18, 104. *See also* gender; misogyny
sexuality, 12, 53–54, 60–61, 75
sexual violence, 18
sharks, 70, 79–80

190 INDEX

shell, 1, 52, 81, 130; of oysters, 54–55, 57, 167n4; pearl, 124. *See also* bivalves
shellfish, 53, 57, 120, 122, 153
Shetland, 115, 118
shipping, 1, 32, 87, 90
ships, 40, 104, 139; cargo, 39; factory, 119
Shiva, Vadana, 106
shorelines, 53, 102
shrimp, 25, 131, 153; farming, 5, 9. *See also* prawns
"silver darlings," 115–16, 118
"simplified sea," 10, 24, 51
Singapore, 64
Singer, Peter, 25
sirens, 102–3. *See also* mermaids
Skojarev, Semi, 94
slavery, 18, 31, 40, 115, 146
Slow Fish, 143
small lines, 114
social science, 17–18, 31, 105, 125
South America, 83
South Australia, 6, 12, 19, 61, 165n1; bluefin tuna of, 4, 62, 68–69, 75, 79, 87–89; oysters of, 161
soy, 149, 156
Spencer Gulf, 68, 81
Spillers, Hortense, 109
Spinoza, 14–15, 46
spotter planes, 91–92
starfish, 54–55, 65, 130
Stehr, Hagen, 88, 94–95, 98
Steinbeck, John, 130–31
Steinberg, Phillip, 16, 39–40
stingray, 109
storms, 20, 53, 81
storm water, 2
Stott, Rebecca, 49, 58, 102
Strathern, Marilyn, 52
strikes, 119
subjectivity, 36, 38, 43, 52, 110
suicide, 6, 18
super foods, 133
supplements, 5, 130, 134, 138. *See also* fish oil
survey, 27
sushi, 78, 85, 87, 96–97, 168n3
sustainability: campaigns for, 29; choices of, 28; complexity of, 162; corporate schemes

for, 28, 33, 166n4; gender and, 104–6, 108; human-tuna-fisher, 94; models, 7–9, 46–47, 147; ocean, 39; problems, 31, 34, 46; seafood, 32. *See also* ecofeminism; Food and Agriculture Organization (FAO); integrated marine trophic aquaculture (IMTA)
swimming, 12, 32, 75, 77–83, 97, 100
swordfish, 32, 137
Sydney, 2, 59, 71, 95, 145, 155, 159
Sydney Cove, 59
Sydney Fish Market, 159–62
Sydney Harbour, 1–2, 4, 14, 59
Sydney Opera House, 52
Sydney rock oysters, 1, 51, 59
symbiosis, 67, 107, 130

Tasmania, 19, 24, 70
taste, 77; acquired, 35–36; anchovies and, 145, 147; bluefin tuna and, 15, 85–87, 91, 96; bodies and, 47; Bourdieu on, 58; identity and, 59; memory of, 49, 152; and oysters, 12, 48, 50, 52, 61, 64, 68, 70–71, 74; relations of, 59–60, 75. *See also* class (social); food
tattoo, 62
techne, 42, 52
technology: bluefin tuna and, 85; and fish industry, 5, 7, 42, 91–92, 114, 119, 153; food, 70, 149; freezing, 4, 90. *See also* integrated marine trophic aquaculture (IMTA)
testosterone, 53
Thailand, 9, 146, 150, 161
Thalassa, 113
Thames (London), 57
Thieme, Marianne. See *Sea the Truth*
ThisFish, 4
Thompson, Paul, 117–19, 126
Thrift, Nigel, 16
thronging, 133
Throsby, Karen, 79
tilapia, 20
Times, The, 58
Tokyo, 85, 87, 91; Tsukiji Fish Market, 19, 78, 83, 85–86, 96. *See also* Japan
Tony's Tuna International, 88, 97
tourism, 1, 39, 73, 80, 159

toxic, 2, 15, 67, 139
toxins, 32, 138
tracking of fish, 4. *See also* spotter planes
trade: Aboriginal, 124; endangered species, 96; herring, 117, 119; illegal, 146; routes, 5, 157; ships, 39; tuna, 86–88, 96
transcorporeality, 33
trawlers, 44, 81, 122, 138, 144; illegal, 31; mega-trawlers, 96; supertrawlers, 120, 138, 157
trevally, 71
trophic cascading, 24
tropics, 1, 26
Tsing, Anna, 40, 126, 130
Tsukiji Fish Market, 19, 78, 83, 85–86, 96
tsunami, 27
tuna, 32, 138; canned, 133; endangered, 78; farming of, 80; fishing for, 81–91, 95; industry, 86, 98; licenses, 124; mercury levels in, 137. *See also* bluefin tuna
Turner Aquaculture, 70

Ulladulla, 99, 167n2
Ulysses, 71
unemployment, 34, 63, 122–23; overcoming, 74
UNESCO, 7, 150
United Kingdom, 9, 64, 71, 138, 149, 166n4. *See also* England; Scotland
United Nations, 8, 39, 93
United States of America, 3, 4, 85, 87, 133, 138–39. *See also* North America
USSR, 119
UV light, 74

veganism, 3. *See also* pescetarianism
vegetarianism. *See* pescetarianism; veganism
village, 34, 114, 160
Virgin. *See* Branson, Richard; *Mission: Save the Ocean*
virus, 149. *See also* disease
vitamins, 53, 138

Waitrose, 27, 109
"Walrus and the Carpenter, The," 55–57
water, 41; deep, 141; oysters and quality of, 15, 53–54, 65, 75; seawater, 24, 37, 65, 140; settlements and, 61; shallow, 62, 84
water pollution. *See* pollution
Waters, Sarah, 54, 58
waves. *See* clapotis
We Are Here, 75
Weheliye, Alexander, 109
Weld, Charles Richard, 19, 113
welfare benefits, 124
West Africa, 96, 138, 157
"wet ontology," 16
Whatmore, Sarah, 110, 168n2
When Species Meet, 110
white death, 3
whitefish, 119–20
Wick, 118
Williams, Meryl, 125, 150
Williams, Raymond, 14, 61
Winberg, Pia, 155–56
Winfrey, Oprah, 133
Winkel, Dos. *See Sea the Truth*
women. *See* feminism; fishwives; fishwomen; gender
Women's Industry Network Seafood Community, 18
wool, 63
working class. *See* class (social)
World Aquaculture Congress, 152–53
World Bank, 106, 138
WorldFish, 125
World Trade Organization, 124
World War I, 39, 119, 133
World War II, 12, 86, 119, 133, 142, 162
World Wildlife Foundation (WWF), 28, 33, 47, 78, 134–35
worms, 130, 136; sea, 114

Yusoff, Kathryn, 18, 20

zinc, 51, 53, 59, 75
zoology, 54, 103, 143
zooplankton, 52, 137